SpringerBriefs in Geography

SpringerBriefs in Geography presents concise summaries of cutting-edge research and practical applications across the fields of physical, environmental and human geography. It publishes compact refereed monographs under the editorial supervision of an international advisory board with the aim to publish 8 to 12 weeks after acceptance. Volumes are compact, 50 to 125 pages, with a clear focus. The series covers a range of content from professional to academic such as: timely reports of state-of-the art analytical techniques, bridges between new research results, snapshots of hot and/or emerging topics, elaborated thesis, literature reviews, and in-depth case studies.

The scope of the series spans the entire field of geography, with a view to significantly advance research. The character of the series is international and multidisciplinary and will include research areas such as: GIS/cartography, remote sensing, geographical education, geospatial analysis, techniques and modeling, landscape/regional and urban planning, economic geography, housing and the built environment, and quantitative geography. Volumes in this series may analyze past, present and/or future trends, as well as their determinants and consequences. Both solicited and unsolicited manuscripts are considered for publication in this series.

SpringerBriefs in Geography will be of interest to a wide range of individuals with interests in physical, environmental and human geography as well as for researchers from allied disciplines.

More information about this series at http://www.springer.com/series/10050

Elena Pede

Planning for Resilience

New Paths for Managing Uncertainty

Springer

Elena Pede
Interuniversity Department of Regional
and Urban Studies and Planning
Politecnico di Torino
Turin, Italy

ISSN 2211-4165 ISSN 2211-4173 (electronic)
SpringerBriefs in Geography
ISBN 978-3-030-17261-9 ISBN 978-3-030-17262-6 (eBook)
https://doi.org/10.1007/978-3-030-17262-6

This Springer imprint is published by the registered company Springer Nature Switzerland AG
The registered company address is: Gewerbestrasse 11, 6330 Cham, Switzerland

Foreword

More than twenty-five years have passed since the publication of Ulrich Beck's famous *Risk Society*, and the role that risk and uncertainty play in our society is now indisputable. Both the frequency of hazardous processes and disasters and the magnitude of their resulting damages are increasing due to natural forces and human activities in mutual interaction.

Given this context, the success of the notion of resilience should not surprise. The consequences of risks related to climate change, financial crises, terrorist attacks, etc., are more and more difficult to prevent completely; the ability of socio-ecological systems to absorb these disturbances—or, even better, to self-organise in new, stable and more adapted states of equilibrium—therefore becomes crucial.

In the last decade, the concept of resilience has been the subject of theoretical redefinitions in many disciplines, according to a complex and co-evolutive perspective. This even at the cost of often making it a fuzzy, umbrella term, which may have limited heuristic significance.

Planning for resilience has the merit to highlight some conditions and approaches that allow to better operationalise the concept of resilience, in the field of urban and regional planning. One of the key goals of this discipline has always been anticipating and responding to disasters; but considering this task through the lens of resilience entails a shift from a "command-and-control" to a "learn-and-adapt" approach in planning.

Therefore, the book focuses on two main issues that are fundamental to pursue more resilient cities and territories through dynamic, fluid and interpretive planning: governance and knowledge. Risk management requires the involvement of a plurality of actors and institutions, which operate in different sectors and on various scales. At the same time, diverse types and sources of—both explicit and tacit—knowledge have to be integrated in order to produce a common understanding for facing risks.

Elena Pede points out how governance and knowledge can mutually reinforce. The co-production of knowledge through collaborative learning between experts and users can help create unified knowledge systems, but this requires governance structures in which hierarchical and bureaucratic interrelations are not barriers to social learning. Intelligent technologies, and in particular geographical information systems, are suggested as a resource to support this process, as they allow to integrate and share spatial data among stakeholders in a collaborative framework.

The experience of the Italian region Umbria in facing seismic risk is finally analysed to show how this approach can work in practice. This case study profitably exemplifies how an organisational innovation aimed at improving knowledge management can enhance vertical and horizontal governance, as well as increase the adaptive capacity of spatial planning.

This book does not downplay or neglect the "dark side" of resilience, in particular its controversial relation with neoliberalism and governmentality forms. At the same time, it shows us how an approach focused on resilience can positively innovate risk management and spatial planning towards a more flexible and efficient perspective in facing the uncertainty and complexity of present and future socio-ecological challenges.

Turin, Italy

Luca Staricco
Associate Professor at Interuniversity
Department of Regional and Urban
Studies and Planning
Politecnico di Torino

Acknowledgements

The publication of this research was supported by the Interuniversity Department of Regional and Urban Studies and Planning (DIST) of Politecnico di Torino. Special thanks should be given to Profs. Luca Staricco (Politecnico di Torino) and Åke Sivertun (Swedish Defence University of Stockholm), my research supervisors, for their professional guidance, encouragement and useful critiques of this research work.

I would also like to extend my gratitude to my friend and colleague Nadia for her suggestions, recommendations and patience.

Finally, I wish to thank my parents, my sister and Marco for their support and encouragement.

Contents

Chapter 1
Introduction: Planning for Resilience

Abstract The Chapter points out the proposal of the book. Over the last decades, the society is becoming more interdependent and more complex, and consequently, more vulnerable to new threats and risks associated. In this scenario, the resilience thinking is gaining prominence in different fields as an attractive perspective for managing uncertainty. However, the application of resilience is hard to be operationalized. The book explores how resilience implies a challenge of flexibility and dynamism in the knowledge systems and learning processes. This chapter presents the structure, the key concepts and the contents of the book.

Keywords Resilience · Uncertainty · Planning for resilience · Flexibility · Knowledge

Our modern society is becoming ever more interdependent and more complex, and consequently more vulnerable to threats and risks. The rapid urbanisation, occupation of unsafe environments, as well as the increased rate of occurrence of climate events are introducing new elements of uncertainty. Likewise, the unprecedented connectivity of the globalised world contributes to increasing interdependencies, and consequently, the risks associated.

Environmental, economic, social, and political problems no longer seem manageable by traditional government processes, and the idea of security that was crucial to risk management in the past has collapsed. The increasing uncertainty and the emerging of new types of risks require alternative paths of planning practice that acknowledge and interact with the implications of the risk society (Beck 1992, 2006, 2009).

In recent years, resilience is increasingly becoming a popular term within policies, programming and thinking as a means of capturing complexity. The concept of resilience was early adopted in ecologic studies during the 1960s, but resilience thinking rapidly came to influence many other research fields with various interpretations, ambiguities and a very wide extension of meaning.

Generally, resilience is the capacity of a system to deal with changes and continue to develop. In the original idea, it referred to dynamics close to equilibrium and, therefore, is defined as the time required for a system to return to an equilibrium point following a disturbance event (engineering resilience) (Holling 1973, 1996).

E. Pede, *Planning for Resilience*, SpringerBriefs in Geography,
https://doi.org/10.1007/978-3-030-17262-6_1

In the socio-ecological system, it refers to dynamics far from any equilibrium steady state and is defined as the capacity of a system to adapt, innovate and transform certain conditions before changing to another stable regime (Holling 1996; Gunderson and Holling 2002; Walker et al. 2004; Gunderson et al. 2009).

Over the years, resilience has emerged as an attractive perspective, especially, with regarding complex systems associated with civil contingencies and crisis management (Godschalk 2003); however, its meaning is far from clear (Restemeyer et al. 2018). On one hand, resilience is something desirable, to aim at; on the other, its fuzziness makes resilience difficult to measure and to be operationalised (Chandler and Coaffee 2016).

The intent of this book is to investigate the principal issues related to the operationalisation of resilience. To do so, the book analyses—regarding emergency planning—the decision-making process by highlighting scale issues in governance levels and in knowledge, multiple perspectives, uncertainty and self-organisation.

1.1 The Book Structure

The book is organised in six chapters. The first chapter is an introduction to the book structure. The second chapter introduces the reader into the complexity of the world where we now live and the changing promises of security. The first two paragraphs examine the increase of pressure on territories, the nature of contemporary risks and the challenges they have created. Subsequently, the chapter problematises the perspective of risk as unknowable or unpredictable and the consequences on policy (Sect. 2.3); while the final paragraph looks at the repercussions of the ongoing changes on the debate in social science.

The concept of resilience as an approach to uncertainty is explored in the following chapter. Sect. 3.1 introduces the conception of resilience, while Sects. 3.2 and 3.3 analyse its diffusion. Key issues of the chapter are the implication of the social-ecological paradigm of resilience for complex systems and for planning. Finally, one of the paragraphs (Sect. 3.4) also investigates the major critiques to resilience such as the excess positivism and the dominance of the neoliberal approach.

The fourth chapter explores the gap between the theoretical conceptions of resilience and how it can be utilised in practice in risk cycle management. The Sect. 4.1 provides the conceptual framework to re-think risk management by drawing on resilience. Section 4.2 analyses the prevision-prevention-response-recovery chain that is traditionally followed in risk management, highlighting its linearity and the rigid categorization of actions. Likewise, the multiple levels and sectors of interventions are analysed (Sects. 4.3 and 4.4), as well as the cross-scale challenges to face (Sect. 4.5). More in detail, the chapter raises several issues: the temporal, spatial and jurisdictional scales of risk management, the growing complexity of governance and the lack of coordination due to heterogeneous expertise.

In light of the findings, the next paragraphs suggest the potential of knowledge and social learning for resilience. In particular, they argue the need of more attention to the roles of production, implementation and dissemination of knowledge with a better integration of multiple scales, different actors, various knowledge sources, and diverse disciplines (Sects. 4.6–4.8). They also investigate the application of technologies, the collaborative production of knowledge through learning processes and the possibility to engage crowds to improve resilience (Sects. 4.9–4.11).

Chapter 5 provides the experience of Umbria region, in Italy, as an attempt to operationalise resilience thinking in emergency management and planning practice. The case illustrates how the integration and the coproduction of knowledge can trigger a process of transformation of planning practice towards a resilience prospective. The most difficult challenge of all to face in this process was the barriers in communication among different disciplines which has required a high degree of reflexivity and social learning. The chapter shows the measures implemented, the strengths and the weakness of the process.

Chapter 6 concludes the book by drawing key themes. In doing so, it briefly traces the evolution of the definitions of resilience and outlines the potential of resilience in renewing and broadening the governance in planning.

References

Beck U (1992) Risk society. Towards a new modernity. SAGE Publications Ltd, London

Beck U (2006) Risk society revisited. The risk society and beyond

Beck U (2009) Critical theory of world risk society: a cosmopolitan vision. Constellations 16:3–22

Chandler D, Coaffee J (2016) The Routledge handbook of international resilience. Taylor & Francis

Godschalk DR (2003) Urban hazard mitigation: creating resilient cities. Nat Hazards Rev 4:136–143

Gunderson LH, Holling CS (2002) Panarchy: understanding transformations in systems of humans and nature. Island Press, Washington

Gunderson LH, Allen CR, Holling CS (2009) Foundations of ecological resilience. Island Press, Washington

Holling CS (1973) Resilience and stability of ecological systems. Annu Rev Ecol Syst 4:1–23

Holling CS (1996) Engineering resilience versus ecological resilience. Engineering within ecological constraints. National Academy Press, Washington, DC, USA, pp 31–34

Restemeyer B, van den Brink M, Woltjer J (2018) Resilience unpacked–framing of 'uncertainty'and 'adaptability'in long-term flood risk management strategies for London and Rotterdam. European planning studies, pp 1–21

Walker B, Holling CS, Carpenter SR, Kinzig A (2004) Resilience, adaptability and transformability in social–ecological systems. Ecol Soc 9:5

Chapter 2
The Promise of Security

Abstract The chapter is an introduction to the challenging period we live in, characterised by the constant reminder of the unpredictability of catastrophic climate events, terroristic attacks, economic crisis, or mass failures of infrastructure systems. At the same time, we are witnessing our planet "shrinking" because of events that generate consequences that have global impacts (e.g. Chernobyl, global warming, 9/11 terrorist attacks, the financial crisis of 2008, etc.). The nature of contemporary risks is unprecedented in terms of their spatial, temporal and potential impact and the traditional categories to control and measure risks are no longer valid. In this context, there has been an evolution of the terms risk and disaster that has enhanced the interest of several academic disciplines; in particular, a new debate has challenged the social science tradition. To face hazards that escape the logic of control, the challenge is to find new ways of relating risks to decisions and practices. The different paragraphs explore how disasters have changed over the years, the evolution of the concept of control embodied in planning for risk in dealing with today's challenges and the key issues of the concepts of vulnerability and risk. The last paragraph concludes the chapter by showing the implications of the world risk society introduced by Ulrich Beck.

Keywords Natural disasters · Risk · Vulnerability · Uncertainty · World risk society, disaster risk reduction

2.1 Living with Perils

Disasters have affected people since the beginning of our existence. The term derives from the Greek negative prefix *dis-* ("apart") plus *aster*, which means "bad star" and refers to the idea that a catastrophe could be traced to the negative influence of a star or a planet. The current definition of disaster is no longer linked to the original meaning of the word. The Centre for Research on the Epidemiology of Disaster (CRED) defines a disaster as (Below et al. 2009, p. 16)

> a situation or event which overwhelms local capacity, necessitating a request to a national or international level for external assistance; an unforeseen and often sudden event that causes great damage, destruction and human suffering.

E. Pede, *Planning for Resilience*, SpringerBriefs in Geography,
https://doi.org/10.1007/978-3-030-17262-6_2

Compared to its original meaning, nowadays the main challenge in this field is the rejection of fatalism and the acknowledgement of disaster as a consequence of the inappropriate management of the ecosystems and the risks associated within it.

Today, we are living in a challenging period characterised by acceleration and interconnectedness in every field of human activity. The population growth, the rapid urbanisation[1] and the intensification of agriculture are putting more and more pressure on the natural systems. At the same time, the complexity of the world may make use more vulnerable than ever. Globalisation has introduced new patterns of exposure and vulnerability, increasing the speed and the consequences of the impacts. For instance, the financial crisis began in a segment of the US mortgage market, but within months it affected most of the countries. In the same way, mosquito populations that are frequently found in tropical areas may easily spread even into Europe due to global warming. Paradoxically, in recent years the rate of disasters and their resulting damages have increased regardless of human progress (Smith 2009).

Disasters[2] can have many forms and can be categorized in different ways. The main distinction is based on the presumed cause: the force of nature, human action, or, in some cases, by the combination of both natural forces and human activities (Vale and Campanella 2005). The latter is increasing; in fact, human actions are contributing more and more to hazardous processes due to the increasing interconnection of nature and society. In other words, all disasters can be seen as being man made, because we cannot prevent hazards from happening, but we can avoid them from developing into a disaster. In addition, man-made disasters can also produce cascading effects; for example, in 1999, the sinking of the Erika tanker off the coast of Brittany caused the release of thousands of tons of oil into the sea, which devastated marine life.

---

[1]The urban population in 2009 accounted for more than half of the total global population.

[2]The terms disaster, hazard, risk and vulnerability will be deeper explored in Sect. 2.3. However, basic definitions on these concepts developed by UNISDR to promote a common understanding on the subject for use by the public, authorities and practitioners are provided herein. Some of these terminologies will also return further on.

> Disaster. A serious disruption of the functioning of a community or a society involving widespread human, material, economic or environmental losses and impacts, which exceeds the ability of the affected community or society to cope using its own resources. Disasters are often described as a result of the combination of: the exposure to a hazard; the conditions of vulnerability that are present and the insufficient capacity or measures to reduce or cope with the potential negative consequences.
>
> Hazard. A dangerous phenomenon, substance, human activity, or condition that may cause loss of life, injury or other health impacts, property damage, loss of livelihoods and services, social and economic disruption, or environmental damage.
>
> Risk. The combination of the probability of an event and its negative consequences.
>
> Vulnerability. The characteristics and circumstances of a community, system or asset that make it susceptible to the damaging effects of a hazard (http://www.unisdr.org/we/inform/terminology).

The fear of man-made disasters rose during the Cold War and has been increasing since Chernobyl; however, the September 11 attacks have strengthened the risk perception and the public attention to man-made disaster in the issue that is becoming more salient today than before (Birkland 2004).

Likewise, disasters can be classified both as rapid-onset and slow-onset (or creeping) (UNDP 2004). The former is defined by instantaneous shocks (e.g. flash-floods, volcanic eruptions, earthquakes, tsunamis, tornadoes, etc.); while the latter may stand for many days, months or even years (e.g. droughts, climate change, desertification, soil erosion, and the AIDS epidemic, etc.).

The CRED classified natural disasters into five major groups (Below et al. 2009, p. 16):

1. Geophysical disasters: events originating from solid earth (e.g. earthquake, volcanic eruptions, landslide, avalanche and subsidence);
2. Meteorological disasters: events caused by short-lived (small to mesoscale) atmospheric processes as in the spectrum from minutes to days (e.g. storm, hurricane);
3. Hydrological disasters: events caused by deviations in the normal water cycle and/or overflow of bodies of water caused by wind setup (e.g. flood, flash-flood, wet mass movement);
4. Climatological disasters: events caused by long-lived/meso to macroscale processes as in the spectrum from intra-seasonal to multi-decadal climate variability (e.g. extreme temperature, drought, wildfire);
5. Biological disasters: disasters caused by the exposure of living organisms to germs and toxic substances (e.g. epidemic, insect infestation and animal stampede).

Similarly, man-made disasters are those events caused directly and principally by one or more identifiable, deliberate or negligent human action, which are classified into two major groups:

1. Technological disasters: disasters due to engineering failures, transport disasters, and environmental disasters (e.g. transport accidents, big fire, chemical contamination, nuclear accidents);
2. Sociological disasters: criminal acts, terroristic attacks, riots, war, stampedes, etc.

Most disasters, whether natural or man-made, can be considered to be a consequence of development failure. For this reason, disasters also affect economic and social development because the time and resources indispensable for remediation impair the stability and development of communities. These are commonly called *secondary effects* and involve short and long-term impacts of a disaster on the overall economy and socio-economic condition (UNDP 2004). They are evaluated both in terms of human and economic impacts. Human impact refers to the number of victims which comprehends the number of deaths[3]; the amount of people affected

[3]Generally, people killed includes those confirmed dead and those missing and presumed dead.

and requiring immediate assistance during an emergency; the total amount of people who report injuries or are homeless. Likewise, the economic impact consists in direct and indirect consequences on the local economy. The shock of a severe disaster can create a cascade of disruption among interdependent systems, shattering the functional capacity of the area (Birkland 2004). The direct consequences concern the damage to infrastructures (electric power, communications, transportation, water, gas and sewage distribution), building and harvest losses; conversely, indirect effects refer to market destabilisation, unemployment, and the loss of revenues. Consequently, the operational capacity of a complex system can have dysfunctions and can be affected by new danger for the population. In the same way, disastrous events can aggravate other stress such as economic weaknesses, social/political conflicts, spreading of diseases and environmental degradation (Jha 2010). For instance, droughts or earthquakes usually have an impact over time and it takes years to get back to normality. For all these reasons, social-economic consequences are a salient issue of post-disaster recovery due to the difficulty to overcome them.

In 2016, the Annual Disaster Statistical Review (CRED 2017) reported that the number of natural disasters during the year was less than the average annual frequency observed from 2006 to 2015 (342 rather than 376.4). Fewer disasters have corresponded to a decrease in deaths, however, the number of people affected, and the economic damages reveal an increase to above average levels. It has been estimated that the number of people affected by natural disaster amounted to 1.5 times the annual average (564.4 million rather than 224 million) in 2016, resulting in the highest costs since 2006 and the economic damages have been placed as the fourth most costly since 2006 (US $ 154 billion). Asia has been the continent most often hit by natural disaster over the last decade, followed by the Americas and Europe, but all continents have been strongly affected. Usually, poor countries register greater human impacts, while economic impacts are felt more in rich countries. Table 2.1 compares the top 10 natural disasters in 2016 in terms of deaths, people affected and economic damages. The former and the second groups are mostly composed by less developed countries (LDCs), while the ranking of natural disasters by economic damages concern mainly more developed countries (MDCs). However, it must be considered that if we compare the economic damages to the countries' Gross Domestic Product (GDP), the impact would be higher with more long-term consequences in poor countries (CRED 2017).

In addition to human and economic consequences, psychological injuries can be also profound. This impact is hard to measure since it is not necessarily proportional to the scale of a disaster and because trauma can persist long after the physical impact of a disaster. For instance, the trauma of 9/11 terrorist attacks hit all the Western World and enhanced the sense of fear and the risk perception (Vale and Campanella 2005).

The book focuses mainly on natural disasters, mostly due to the need to restrict the research field. Specifically, each catastrophic event has different actors involved according to the type of disaster, and to the spatial, temporal and jurisdictional extension; for these reasons it is impossible to talk about a unique methodology of action for natural disasters. However, part of the suggestions that arise can also be applied to other types of catastrophe, as for instance, the multi-dimension of territorial

Table 2.1 Comparison among the top 10 natural disaster in 2016 by number of deaths, by the total of people affected and by economic damages

Top 10 natural disasters in 2016					
By number of deaths		By total people affected		By economic damages	
Country	No. deaths	Country	Tot. affected	Country	(US$ billion)
China	1,335	India	333,816,813	China	45,328
India	1,142	United States of America (the)	85,123,038	United States of America (the)	44,035
Ecuador	686	China	72,387,822	Japan	20,200
Korea (the Democratic Republic of)	612	Ethiopia	10,696,080	Italy	5,300
Haiti	599	Malawi	6,702,800	India	4,074
Pakistan	409	Haiti	5,794,943	Canada	4,000
Italy	299	Somalia	4,700,000	France	2,400
Zimbabwe	277	Vietnam	4,567,721	Germany	2,000
Indonesia	270	Philippines (the)	4,416,499	Thailand	1,245
United States of America (the)		Zimbabwe	4,302,128	Argentina	1,000

Source Annual Disaster Statistical Review 2016 (CRED 2017)

resilience and the sharing of knowledge in order to improve capability in managing emergencies.

2.2 Who Is Responsible for Natural Disasters?

The awareness of hazards and disasters has changed through history. As Quantarelli (2000) affirms, it is possible to recognize three main phases in this field. In the past, when the sources of destruction were largely caused by natural forces, people used to look at disasters as *acts of God*. This perspective emphasized the idea of catastrophes as divine punishments or fate, rather than as the consequence of pressure from human activities on the ecological systems. This fatalistic attitude consequently implied that nothing could be done to prevent disasters and it also paralysed the production of new scientific knowledge for tackling them. The rise of Enlightenment secularism and the development of the scientific method and thinking contributed to increase the awareness of the terms *hazard* and *risk* and their interdependence with human activities. The first great step took place between the 19 and 20th centuries, when science started to consider disasters as *acts of nature*. This implied the prospect of working on preventive actions and impact reduction that firstly concerned only engineering measures. Lastly, at the beginning of the 20th century, two works (Prince 1920; White 1945) led the way to modern hazards and disaster studies exploring the role of human behaviour and decision-making. Despite these first studies on human responsibility, it took half a century for disasters to be considered a result of *acts of society*. In this third paradigm, disasters are considered social consequences resulting from inappropriate actions of human beings (Quarantelli 2000). They occur due to the failure of society that creates vulnerably-built environments that can be impacted (e.g. non-earthquake proof buildings or urban areas in flood plains, etc.).

The historical evolution of the term disaster—from a conceptualisation as a physical agent to one which mostly emphasized social features—has enhanced the interest of several academic disciplines, overcoming the engineering paradigm in this field. In particular, the claim of the Anthropocene narrative of the human force in shaping the functioning of the Earth system established challenges that necessarily require experts, knowledge, theories, approaches, tools and the methods of a variety of academic disciplines (Dominey-Howes 2018). In the same way, the advancement in acknowledging the sources of disasters prompts the question concerning the *responsibility* in taking the risk, in protecting the society, or for what we have not done (Gunder and Hillier 2007). In our contemporary world where risks are characterised by a mix of complexity and uncertainty and are not restricted to national borders, the issue of responsibility should be more debated.

2.3 Risk and Vulnerability

Before going into the content of the book, it is useful to make a brief overview of the concepts of risk and vulnerability and their evolution over time. The definitions presented in this paragraph consider several existing sources but also show the dynamic of these fields and the evolving knowledge over time.

In common acceptance, the concept of risk refers to potential negative (undesirable) consequences that may arise from a future event. Generally, the definition of risk is the likelihood of an event's occurrence multiplied by the severity of the consequences of that event if it occurs.

(a) Risk (R) = Likelihood (L) * Consequence(C)

However, in disaster studies, the most commonly used formulation was defined by UNESCO in 1972. The definition establishes that risk is the combination of three components: Hazard (H), Vulnerability (V) and Exposure (E)

(b) Risk (R) = (H) Hazard * (V) Vulnerability * (E) Exposure

(H) the probability of the occurrence of a potentially damaging phenomenon within a given time period and area.

(V) the degree of loss to a given element or set of elements at risk resulting from the occurrence of a natural phenomenon of a given magnitude.

(E) elements at risk: population, property and economic activity at risk in a given area.

In short, risk concerns impacts (damages, human injuries and deaths) and it is linked to hazard and vulnerability. The term hazard is rooted in the concept of change (Jha 2010) and denotes a danger or a potentially harmful situation. Hazard is an intrinsic factor or a latent condition that has the potential to damage, harm or create losses. According to UNISDR hazard is defined as

> A potentially damaging physical event, phenomenon or human activity that may cause the loss of life or injury, property damage, social and economic disruption or environmental degradation (2004, p. 4)

and it is associated with a level of risk and the degree of vulnerability of the system. A hazard per se in not a disaster but a disaster is a social happening, where different elements can create cascading scenarios. According to the research fields of application, several analysis methods are used. For example, in the last decade there has been an acceleration in the use of a systemic or an all hazard approach, rather than setting up agent specific entities or functions (e.g. for floods or chemical threats).

As already argued, at the end of the XIX century, weather forecasting scientific tools and civil structures with the scope of defence were widely diffused since the main issue of scientific research was how to protect humans, their activities and buildings from damaging consequences. During this period, the interactions between environmental hazards and human behaviour were underrated; only in the mid-twentieth century did the behavioural paradigm start to be explored by introducing the role of people in minimising risk.[4]

Since the 1990s, the emphasis on mutual interactions between nature and society has grown; the complicated mix of interaction between human and natural systems has increased, especially the factors of vulnerability, showing the complexity of the causes of disaster. More in detail, vulnerability refers to the capacity of a system to absorb or cope with the consequences of external effects. The Intergovernmental Panel on Climate Change since its Second Assessment Report (1995) has declared that vulnerability is the function of the sensitivity of the system to change, of its adaptive capacity and of the degree of exposure of the system to hazards. It means that a system is more vulnerable when modest changes produce harmful consequences. The concept has been developed since the 1970s and subsequently modified in the different fields. Over the years the term has conserved as key parameters

> the stress to which a system is exposed, its sensitivity and its adaptive capacity (Adger 2006, p. 269)

but it has also enriched by taking into account more factors such as Eakin and Luers (2006):

---

[4]The literature related to the influence of human behaviour in the 1970s contemplated a more radical alternative that focused on the differences between more MDCs and LDCs. This new paradigm put the development at the centre of its research, in order to investigate in which way a less developed economy contributes to vulnerability.

- the interactions among stressors;
- the socioeconomic and biophysical framework;
- the role of equity and accessibility to resources;
- nested scales and scalar dynamics of hazard.

For this reason, it is possible to identify two types of vulnerability: territorial (physical) and systemic vulnerability. The former refers to spatial factors and characteristics; the latter, instead, indicates the relational characteristics between human and natural systems.

The literature on vulnerability is characterised by a strong multidisciplinary approach, fragmentation and lack of dominant paradigms. However, it is possible to outline three main lines of research (Eakin and Luers 2006):

1. the risk-hazard models: these studies focus on mitigation where vulnerability is equated with the analysis of stressors and the potential impacts;
2. the pressure and release models: it concerns studies in political-economic and political-ecological fields, where vulnerability and adaptive capacity are dependent on conditions such as accessibility and fairness in the distribution of opportunities and resources;
3. ecological resilience studies: risk and vulnerability are interpreted as dynamic properties of a system where social and environmental dimensions are constantly changing.

The risk-hazard model has evolved from the natural hazard literature in geography; the main focuses are the identification of external events and the location where crises may occur as well as the socio-economic consequences that might be expected (McCarthy et al. 2001; Cutter and Finch 2008). In this model, vulnerability is the function of the exposure to hazardous events and the sensitivity of the exposed entity (Turner et al. 2003), conditions that make people or places vulnerable to hazards. In this approach the evaluation of vulnerability is an ex post identification and it works from hazard to impacts. This model does not take into account the ways the systems amplify or attenuate the impacts of the hazard and the role of social structure and institutions in managing exposure and consequences (Turner et al. 2003).

In the political-economic approach (PAR), the system exposure to risk and the adaptation ability to unforeseen consequences is based on social, economic, political, cultural and biophysical factors (Pelling 1999; Rose and Liao 2005). Risk is explicitly defined as a function of perturbations, stressors and vulnerability of the exposed entities. It is characterised by the interaction of scales of causation and the social differences, as well as by the emphasis on historical processes (Eakin and Luers 2006). The model identifies three categories of components on the social side: root causes, dynamic pressures and unsafe conditions and one component on the natural side, the natural hazard itself. The inequities in resource access and opportunities, the social marginalisation and the institutional framework determinate the state of vulnerability. Despite the emphasis placed on "social" conditions of exposure,[5] the

[5]In this model the livelihood can be defined by drawing upon and combining five types of *capital* (Wisner 2004, p. 96):

PAR model does not take the vulnerability of biophysical subsystems (Wisner 2004) enough into account.

The last line of research considers vulnerability as a dynamic property of a system where human and ecological systems have a continuous interaction. The resilience paradigm focuses on the evolutionary trajectories of a system according to its ability to absorb changes and disturbances (Holling 1973). Resilience theory was firstly applied to ecological systems than to social sciences over time. In this approach, resilience and vulnerability have their point of contact in the need to change the response of systems and the role of actors.

Despite the traditional negative conception of vulnerability, the resilient paradigm emphasizes the capability of the system to absorb disturbances as well as its capacity to self-organise and to evolve in a new stable-state (Adger 2006).

Finally, Turner et al. (2003) argued that vulnerability concerns the totality of the system that consists of:

- The components of exposure;
- The human-environment condition of the system that determinates its sensitivity to any set of exposures;
- Coping, impacts, adjustments, and adaptations. These elements are scale dependent and interactive.

2.3.1 *The Uncertainty of Risk in the World Risk Society*

Commonly, planning for risk embodies the concept of control, which presumes a decision-making process based on the idea that risk can be measured, observed and mapped in order to be governed. Under this perspective, planning has often been perceived as a scientific process where decision-makers must anticipate what could happen in the future.

Since the 1990s, a new perspective of risk as unknowable or unpredictable has come to light. The increase of population density and urbanisation, the expansion of inter-connectiveness between natural, technical, social and economic risks and the growing uncertainty about natural hazard patterns and frequencies due to the global change have made unprecedented challenges arise. The contemporary society have to face risks as far ranging as degradation to the global ecology, international terrorism, global health pandemics, or the health consequences feared as a result of exposure to a myriad of technologies, chemicals, industrial toxins and pollutants and financial crisis (Jarvis 2007).

---

1. Human capital (skills, knowledge, health and energy);
2. Social capital (networks, groups, institutions);
3. Physical capital (infrastructure, technology and equipment);
4. Financial capital (savings, credit);
5. Natural capital (natural resources, land, water, fauna and flora).

This changing nature of risk and the different attitudes towards risk triggered the debate that has changed the social science tradition.

The best known approach is the '*risk society*' perspective introduced by Ulrich Beck in 1992. One of the main issues of risk society refers to the concept of risk. In his work, Beck argued that the speeding up of modernisation has produced a "watershed" between the world of quantifiable risk, in which decision-makers can think and act based on calculating risks, and the world of non-quantifiable insecurities that we are creating. Modern societies are confronted with the principles and limits of their own development model because the western modernisation and industrialisation has increased the complexity of risk in the contemporary society. Risks are generated by society and not by nature. The concept of risk society takes this as its starting point, in order to articulate systemic and epochal transformation. It particularly argues a new challenge in the relationship of society to hazards and problems produced by it, which in turn exceed the societal conceptions of security.

According to Beck, this new period corresponds to a second modernisation—called *reflexive modernity* because it requires a

> self-confrontation with the consequences of risk society which cannot adequately be addressed and overcome in the industrial society system (ibid. p. 28).

This turns traditional assumptions about planning and managing the future into inappropriate and they are overtaken by their failures to secure any calculable future (Adam and Adger 2013). This transformation draws the attention to the concept of uncertainty instead of the risk one. In the *reflective of globalisation* theme, Beck (1992) explains that individuals become aware of the contradictions and limitations of the industrial society. There is not only a "reflexivity matter" but also "boundary" and "voice" matters (Cantelli et al. 2010). The boundary matter argues that the traditional categories of nation-state are no longer valid and the governance of uncertainty and risk involve various sectors and policy fields (*cosmopolitanising of risk*) (Beck 2009). On the other hand, the voice matter concerns citizens' participation and inclusion. Nowadays, new roles and identities for citizens, experts and bureaucrats have evolved and multiple types of knowledge beyond scientific expert knowledge have emerged and serve as an ongoing resonance body for risk perceptions and governance requirements.

In this framework characterised by uncertainty, the challenge is to find ways of relating risks to decisions and practices. Contemporary risks are unprecedented in terms of impacts and the possibility of their prevention. Beck emphasises that risk implies decision-making, but in the face of the unknown factor of crisis potential, decisions are increasingly disabled. The emerging of new types of risk requires an alternative mode of planning practice that acknowledges the new conception of security to identify, enable and mobilise the physical, intellectual, social and cultural elements of society. Nowadays, in the context of a world with increasing uncertainty, there is the need to bring a new dimension to the long-standing role of planning in risk management, by shifting the discussion from managing *calculated risks* to *uncalculated risks*.

References

Adams H, Adger WN (2013) The contribution of ecosystem services to place utility as a determinant of migration decision-making. Environ Res Lett 8:015006. https://doi.org/10.1088/1748-9326/8/1/015006
Adger WN (2006) Vulnerability. Glob Environ Change 16:268–281
Beck U (1992) Risk society. Towards a new modernity. SAGE Publications Ltd, London
Beck U (2009) Critical theory of world risk society: a cosmopolitan vision. Constellations 16:3–22
Below R, Wirtz A, Guha-Sapir D (2009) Disaster category classification and peril terminology for operational purposes. CRED, Brussels
Birkland TA (2004) Introduction risk, disaster, and policy in the 21st century. Am Behav Sci 48:275–280. https://doi.org/10.1177/0002764204268997
Cantelli F, Kodate N, Krieger K (2010) Questioning world risk society: three challenges for research on the governance of uncertainty. Glob Policy
CRED (2017) Annual disaster statistical review 2016 - the numbers and trends. Center for research on the epidemiology of disasters (CRED), Brussels
Cutter SL, Finch C (2008) Temporal and spatial changes in social vulnerability to natural hazards. Proc Natl Acad Sci 105:2301–2306
Dominey-Howes D (2018) Hazards and disasters in the Anthropocene: some critical reflections for the future. Geosci Lett 5:7
Eakin H, Luers AL (2006) Assessing the vulnerability of social-environmental systems. Annu Rev Env Resour 31:365–394
Gunder M, Hillier J (2007) Problematising responsibility in planning theory and practice: on seeing the middle of the string? Prog Plan 68:57–96
Holling CS (1973) Resilience and stability of ecological systems. Annu Rev Ecol Syst 4(1):1–23
IPCC (1995) IPCC second assessment - climate change 1995
Jarvis DS (2007) Risk, globalisation and the state: a critical appraisal of Ulrich Beck and the world risk society thesis. Glob Soc 21:23–46
Jha MK (2010) Natural and anthropogenic disasters – vulnerability, preparedness and mitigation. Springer, Berlin
McCarthy JJ, Canziani OF, Leary NA, et al (2001) Climate change 2001: impacts, adaptation, and vulnerability: contribution of working group II to the third assessment report of the intergovernmental panel on climate change. Cambridge University Press, Cambridge
Pelling M (1999) The political ecology of flood hazard in urban Guyana. Geoforum 30:249–261
Prince SH (1920) Catastrophe and social change, based upon a sociological study of the Halifax disaster. Columbia University, New York
Quarantelli EL (2000) Disaster planning, emergency management and civil protection: the historical development of organized efforts to plan for and to respond to disasters. Preliminary papers 301
Rose A, Liao SY (2005) Modeling regional economic resilience to disasters: a computable general equilibrium analysis of water service disruptions*. J Reg Sci 45:75–112
Smith K (2009) Environmental hazards: assessing risk and reducing disaster, 5 edn. Routledge, New York
Turner BL, Kasperson RE, Matson PA et al (2003) A framework for vulnerability analysis in sustainability science. Proc Natl Acad Sci 100:8074–8079
UNDP (2004) Reducing disaster risk, a challenge for development
UNESCO (1972) Report of consultative meeting of experts on the statistical study of natural hazards and their consequences
UNISDR (2004) Living with risk. A global review of disaster reduction initiatives
Vale LJ, Campanella TJ (2005) The resilient city: how modern cities recover from disaster. Oxford University Press, Oxford
White GF (1945) Human adjustment to floods: a geographical approach to the flood problem in the United States. University of Chicago, Chicago
Wisner B (2004) At risk: natural hazards, people's vulnerability and disasters. Psychology Press

Chapter 3
Resilience: A Critical Background

Abstract The chapter aims to develop a conceptual framework for orienting the resilience-building challenges in planning. In a world in which the only certainty is uncertainty, resilience has become one of the most widely used concepts that has come to prominence in understanding and managing complex systems in the last decade (Welsh, Geogr J, 180:15–26, 2014). The concept of resilience was firstly formulated in ecology during the 1960s, but it has influenced many other research fields. With the advent of the social-ecological perspective, the resilience theory has also influenced political and human geography and environmental studies, with new explorations and paradigms (Davoudi et al, Plan Theory Pract 13:299–333, 2012). In particular, resilience has become a "pillar" of disaster management, climate adaptation, and regional economic development (Davoudi et al, Plan Pract Res, 28:307–322, 2013). However, the widespread use of resilience thinking has not brought any conceptual clarity but has instead contributed to the fuzzy character of the concept. The chapter firstly traces the conception of resilience and its three main currents: engineering resilience, ecological resilience, and socio-ecological resilience. Secondly, it presents how resilience can be linked with vulnerability, and finally it outlines the resilience interpretative approach to emergency management asserted in this work. The chapter also deals with the criticism to resilience in the literature, such as the unified positive meaning given to the concept and the devolution involving responsibility.

Keywords Resilience · Social-ecological systems · Panarchy · Adaptive management · Transformability · Planning

3.1 Engineering and Ecological Approaches

The concept of resilience was borrowed from ecological studies, deriving it from the way ecological systems cope with stress and disturbances caused by external factors. Subsequently, the theory has also been applied to the studies on the adaptive complex systems.

E. Pede, *Planning for Resilience*, SpringerBriefs in Geography,
https://doi.org/10.1007/978-3-030-17262-6_3

The father of the modern resilience thinking is the Canadian theoretical ecologist Crawford Stanley Holling thanks to his studies on population and behavioural ecology. Before his work, the term resilience was drawn into a linear logic denoting resilience as a property of materials. In this form, called *engineering resilience*, the conceptual emphasis was focused on *resistance* and *stability*. Its measurement concerns the qualities inherent to the fabric itself of the object and the magnitude of the disturbance (Adger 2000; Coaffee et al. 2009); it implies that the faster the system bounces back, the more resilient it is Davoudi et al. (2012).

In his studies, Holling actualised resilience by applying it not to materials but to ecosystems and he was the first to introduce the importance of nonlinear dynamics. He defined stability as

> the ability of a system to return to an equilibrium state after a temporary disturbance (Holling 1973, p. 17)

where the steady-state becomes systemic recovery rather than material efficiency. In contrast to engineering resilience, Holling considers resilience in terms of size of stability domains, or more meaningfully, as the ability to absorb and adapt to a disturbance, which is not necessarily the capacity to return to the former steady-state. Consequently, according to Holling

> Resilience determines the persistence of relationships within a system and is a measure of the ability of these systems to absorb changes of state variables, driving variables, and parameters, and still persist. (Holling 1973, p. 17).

This became known as *ecological resilience* (Holling 1996; Adger 2000; Walker et al. 2004). The ecological resilience rejects the existence of a single equilibrium but suggests that there are multiple equilibria. In other words, it is

> the capacity of a system to undergo disturbance and maintain its functions and controls (Gunderson and Holling 2002; Gunderson et al. 2009)

or the buffer capacity to *persist* to external disturbances (Folke 2006).

In summary, engineering resilience focuses on efficiency of function and the vicinity of a stable equilibrium, while ecological resilience focuses on maintaining the existence of functions and multiple equilibria (Holling in Davoudi et al. 2012; Folke 2006). Likewise, engineering resilience applies only to the behaviour of a linear system, where resistance to change is addressed in terms of recovery.

However, reality is different: disturbance events and spatial heterogeneity make recovery trajectory impossible to predict and the system is never the same but it is continuously developing (Folke 2006). For this reason, the concept of alternative stable-states and the ecological resilience perspective were the point of departure for many scholars linked to complex systems and non-linear dynamics (Folke 2006). Complex systems are open systems and because of the second law of thermodynamics, they can change their entropy. This mutation is determined by non-linear interactions between its elements; it means that interactions in complex systems can generate reactions unintuitive and apparently accidental (Bertuglia and Staricco 2000). In complex systems the change of an element can trigger breaking reactions (negative feedback) or reinforce reactions that amplify the initial change (positive feedback).

No-linear interaction means that a local action can have an impact on the whole system because it has a complex behaviour *oriented to stability*. Every time the system overshoots, the steady-state produces a dynamic reaction, after that

> the system may look similar but it is not the same system, because like any living system it is continuously developing (Folke 2006, p. 257).

The result is a continuous process of mutual adaptation and self-organisation of the components that characterises both the biological and socio-economic systems.

Besides, resilience is not only about resisting disturbance and change but also includes all the opportunities that disturbances open in terms of re-organisation and evolution of the structures and processes (Folke 2006). An example would be forest fires that can play an important role in forest ecosystems because they can regulate the underbrush, avoiding more destructive fires. It means that the system has a self-organisation that can clean undergrowth without damaging the older trees through low-temperature fires. In this case, self-organisation improves the ability to respond to external stress adapting towards uncertainty and risk (Bertuglia and Staricco 2000).

However, the idea that the nature of the stability domain does not remain fixed over time was introduced by the socio-ecological approach to resilience with the aim to include the social sphere in ecosystem analysis.

3.2 System Theories: From the General System Theory to the Social-Ecological System

As already argued, resilience is a concept rooted in complexity theory. Before continuing with the analysis of the social-ecological paradigm it is useful to explore the systems theory more deeply.

In the 1930s, the biologist Ludwig von Bertalanffy formulated the *General System Theory* (GST) that was relevant for numerous fields of research. The theory argues that a system is a set of elements that has inter-relationships; consequently, the understanding of the essential properties of the parts is determined by their relations. In this first phase, the model only referred to isolated systems based on the second law of thermodynamics. Afterwards, in 1969, von Bertalanffy implemented the systems theory with the theory of open systems: systems that can also be influenced by external events. In other words, in contrast to isolated systems that contain limited energies, in open systems the energy amount can be altered. In the beginning, the open systems theory was applied only to natural sciences; the concept then also had applications in social sciences. Other scholars, mostly related to the Santa Fe Institute, linked von Bertalanffy's theories to the notion of complex systems (Bertuglia and Staricco 2000). The system's complexity refers to the multiple non-linear interactions and feedback loop and casual chain relations that link system components. It means that phenomena have multiple and dispersed causes that cannot be solved only through scientific solutions organised on traditional disciplinary lines (Berkes et al. 2003). The complexity is related to the system behaviour that is oriented to stability but can reach multiple equilibria. Non-linear interactions between elements mean that most of them are unpredictable and unintuitive and when conditions change there can be positive or negative feedbacks. Through element interaction, systems can self-organise, novel configurations can emerge, and adaptation is made possible. This means that a disturbance can create the challenge for new opportunities for recovering or reorganising (Resilience Alliance 2010). This capacity is related to the maintenance of the diversity and individuality of its components and the localised interactions among those components. A complex adaptive system has four basic properties: aggregation, non-linearity, diversity, and flows (Folke 2006).

Natural and social systems are complex adaptive systems in themselves. Theorising the idea of a new geological epoch called *Anthropocene*, the Nobel Prize winning chemist, Paul Crutzen (2002) argued that today it is impossible to understand nature without society and vice versa (Jahn et al. 2009). In this geological time the influence of human behaviour on the Earth's atmosphere is so significant that human impacts are comparable to those of geological forces (see Chap. 1). Human-nature interactions have increasingly spread the social-ecological system (SES) concept in several research areas with different theoretical backgrounds. It has also contributed to the diffusion of multi—inter—and transdisciplinary researches. For instance, since the fifth Framework Programme, the EU cross-disciplinary objective claims to overcome the traditional structure of research organisations based on sectorial research communities (European Commission 2005).

Nowadays social-ecological systems are the core cognitive interest of different fields of research, especially for sustainable development. More in detail, it is central in resilience thinking because natural resource management issues are the sum of multiple ecological and social integrated elements. SES is the combination of ecological, cultural, political, social, economic, technological components.

Social-ecological systems are complex, adaptive and delimited by spatial or functional boundaries. SES theory was defined the *integration of the parts* (Holling 1996; Glaser et al. 2008) and its main tendencies are:

1. the evolutionary ecological orientation, focusing on adaptive renewal cycles in multi-scale, panarchical structures;
2. the complexity theory, focusing on non-linear dynamic systems and the transfer of system expertise to strategic planning and adaptive management;
3. the quantitative/formal approaches.

As already discussed, systems can change and may shift into a different state; this transition can be slow and gradual but also sudden. Understanding how systems change over time and which are the critical thresholds can potentially provide advance warning or changes to avoid undesirable shifts in system states. Thus, resilience can represent the distance between a system state and a critical threshold.

3.3 The Social-Ecological Paradigm

In the social-ecological approach, the concept of resilience is about the opportunities that emerge in terms of adaptation, learning and self-organisation in addition to the ability to resist disturbances (Berkes and Folke 1998; Folke 2006; Walker et al. 2004).

Social–ecological resilience argues that

> social and ecological systems are themselves linked [...] to synergistic and co-evolutionary relationships (Adger 2000, p. 350).

This means understanding how ecosystems are structured and behave and how institutions and the people associated with them are organised and act. The volume *Barrier and Bridges to the Renewal of Ecosystems and Institutions* (Gunderson et al. 1995) argues that social-ecological resilience is linked to the necessity to learn to manage by change, rather than simply react to it; it can be likened to *bouncing forward* in place of *bouncing back* (Folke 2006; Folke et al. 2010). This is an optimistic view of change that sees the empowerment of agencies but also individuals, small groups or teams of individuals.

In summary, according to Folke (2006) and Gotts (2007), the social-ecological resilience conceptual framework includes:

1. Multiple equilibria: Rather than a single equilibrium point, complex systems generally have multiple metastable regimes;
2. Disturbance:

   > "[..] the magnitude of disturbance that can be absorbed before the system changes its structure by changing the variable and process that control behaviour" (Gunderson and Holling 2002, p. 4);

3. Multiple distinctive scales with cross-scale interactions. Gunderson and Holling (2002) argue that systems form a multilevel hierarchical structure with different degrees of self-organisation.

Therefore, social—ecological resilience is not *a return to normalcy* (Pendall et al. 2010, p. 76) but the ability of complex social-ecological systems to change, adapt or transform into response to stress and strains (Folke et al. 2005). For example, the strong shock caused by hurricane Katrina in New Orleans in 2005, has not only caused severe damages and losses to the system but has also revealed several conditions as unsustainable. Instead of returning to the prior equilibrium, the shock showed the need to reformulate *new order* in terms of society, economy and policies: for instance, some neighbourhoods were not rebuilt because they were placed in unsafe areas.

The multiple distinctive scales with cross-scale interactions is described by Gunderson and Holling (2002) with the concept of *panarchy*,[1] preferred to the idea of

---

[1]The term Panarchy firstly coined by Paul Emile de Puydt in 1860, referring to a specific form of inclusive, universal system of governance that includes all other forms of governance. Panarchy with this acceptation is used in international relations to describe global governance (Edson 2010). At a later stage, the term Panarchy was introduced in systems theory. Holling and Gunderson referred

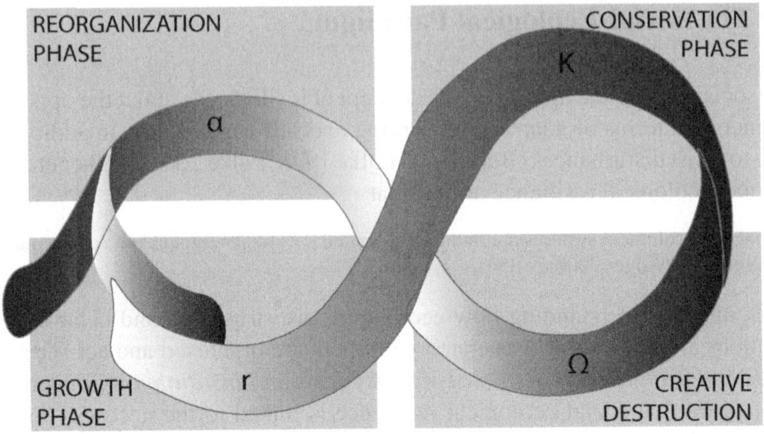

Fig. 3.1 Panarchy, adaptive cycle and change. *Source* elaboration by the author, *primary source* Gunderson and Holling (2002)

hierarchy. Panarchy describes the cyclical and organisational attitude of systems: how an ecosystem organises itself and how it changes and responds to external disturbances and to changes in the world. With the term panarchy, Holling (2004) described the way living systems both persist and innovate at the same time and show how fast or slow, events and processes can transform ecosystems into an evolutionary way, giving changes for learning to their societies.

With the metaphor of the *adaptive cycle* (Fig. 3.1), Gunderson and Holling (2002) suggest that complex systems follow a four-phase cycle of (1) *growth* or *exploration* (r); (2) *conservation* (k); (3) *release* or *creative destruction* (Ω) and (4) *reorganisation* (α). Each phase of the cycle creates the condition for the next phase and every phase entails the loss of resilience and the consequent vulnerability of the system. The cycle is composed by two phases of growth—called forward loop—and two phases of reorganisation (back loop).

Therefore, the adaptive cycle primarily focuses attention on the function of processes and dynamics rather than on states and structure (Pickett et al. 2004). Thus, adaptive cycles and their interactions are not limited to the dynamics of ecosystems, but they similarly occur in societies. This means that individuals, societies, groups, and institutions play a strong role in many ecosystems. Frequently, the influence is in the form of past activities that have a persistent effect on soils, plants, or other major components of ecological systems. Figure 3.1 shows a stylised representation of the four system steps and the flow of events among them. The connections between the different levels of panarchy explain how even small and fast cycles can affect larger

the term to the mythological figure of Greek god Pan, the paradoxical spirit of nature. They joined the idea of Pan to the dynamic reality of hierarchies across scales and to the interactions among them that can influence the phase cycles of one another (Holling 2004). In systems theory the term was coined as an antithesis to the word hierarchy in order to describe the framework of nature's rules.

and slower ones (revolt) or when large and slow ones can control the renewal of smaller and faster ones (memory).

External events can start unpredictable reactions. During such times, uncertainty is high, and control is weakened, but it does not mean that it will result in part of the system collapsing. Indeed, the cross-scale interactions can create space for innovation and reorganisation that means there are opportunities for change and evolution of the system. This concept enlarged the application of panarchy as a framework for different research fields. Panarchy and adaptive cycles are also applied in policy and institution (Ostrom 1990), economic geography (Simmie and Martin 2010), decision-making (Westley 1995) and knowledge systems (Berkes and Folke 1998). If we look at complex natural systems, we can easily find a front-loop phase of slow, incremental growth and accumulation that is more predictable and a back-loop stage of rapid reorganisation leading to renewal or to collapse characterised by a higher uncertainty (Holling 2004). The instability can be part of a process of transformation that can lead to a new phase of opportunity. For a better understanding of cross-scale interactions and their consequences we can look at Arab Spring. The revolutionary wave of demonstrations, riots and civil wars is creating change not only at national levels; since protests began, the flows of immigrants to Europe have changed and economic consequences have been registered by the geopolitical instability in oil mining countries.

The idea of a transforming system has also been interpreted with an evolutionary perspective in planning by Davoudi et al. (2012). In this context, the emphasis is not on the return to normality but—on the contrary—on the research of a new normality (Pendall et al. 2010; Davoudi et al. 2012).

3.4 Connecting Resilience Concept to Planning

Historically, planning is open to absorb new concepts and translate them into its theories and practices (Davoudi et al. 2012). Several terminologies, approaches and metaphors have been rapidly becoming part of planning theories and practices, some of them with solid roots, others with uncritical acceptance—resilience also not being an exception.

During the last decade, as discussed in the previous part, the concept of resilience has increased its scope. It emerged as a fusion of ideas from multidisciplinary traditions e.g. ecosystem, engineering infrastructure, psychology, behavioural sciences and disaster risk reduction. The multiple concepts and approaches to resilience have been widespread in a wide range of research fields also thanks to its fuzzy character. As Keith Shaw highlights

> much of the appeal of the term lies in it being sufficiently malleable (ibid., 2012, p. 308)

which makes it a *versatile umbrella term* (Fünfgeld and McEvoy 2012). It stands to reason that the current context characterised by crisis, uncertainty, complexity and interdependencies may be easily interpreted by the resilience approach that appears to have the potential to play a critical role (Teigão dos Santos and Partidário 2011, p. 1519).

However, several authors have grounds to fear that resilience could be the buzzword of the moment, replacing other commonly used terminologies as sustainability and adaptation (Teigão dos Santos and Partidário 2011; Porter and Davoudi 2012; Davoudi et al. 2013). To contextualise the concept of resilience, it is therefore important to analyse its relationship with sustainability and adaptation and its implications.

Sustainability and sustainable development were born as a concept in 1987 (Brundtland and Khalid 1987), indicating the capability to

> [...]ensure the needs of the present without compromising the ability of future generations to meet their own needs (art.27)

and they rapidly became the most relevant keywords of political action of the contemporary society. The success of the sustainability concept in the past two decades has overwhelmed most research fields, from economy to ecology to the idea of sustainable development as an oxymoron to underline the need of a sustainable de-growth.

Sustainable development focuses on the future and, as Teigão dos Santos and Partidário (2011) argued, it could be considered a precautionary principle that turns its attention to avoid negative consequences of contemporary development in the world of tomorrow. In the same way, the two authors underlined the current increase of instability, crises and pressure on ecosystems that need new and rapid solutions, in a more adaptive way, due to their stronger and faster cascading effects. Therefore, resilience offers something in reaction to the uncertainty and insecurity of contemporary context (Shaw 2012).

In this framework, resilience appears to be a key condition to effectively pursue sustainable development (Lebel et al. in Teigão dos Santos and Partidário 2011). Somehow, resilience could be considered as included in the concept of sustainability, where sustainability is the ability to maintain the functionality of the system when it is disturbed.

Specifically, resilience and its cross-scale interaction play a fundamental role in managing a transition towards more sustainable development. Several authors (Gunderson and Holling 2002; Lambin 2005) recognised that the organisational attitude of the systems in the resilience approach is a fundamental issue of sustainability. In an uncertain and unpredictable reality, looking at social-ecological systems through the lens of resilience enables the consideration of a world where the capacity of its multi-stakeholders' (such as scientists, policy-makers, practitioners, private entrepreneurs and citizens) interaction across scales is fundamental in determining the dynamics of the system. Thus, resilience implies a change in the way of policy-making, planning and governance, it provides a common language across different sectors and discipline (Wilkinson 2012). Resilience implies different assumptions and approaches, meaning a change in processes that go from being essentially rationalist, reactionary or bureaucratic, to becoming more adaptable, anticipative, flexible, collaborative and co-accountable.

In the same way, the issues that link resilience to sustainable development are also in relation to the adaptation concept. Translated into the context of climate change, adaptation can be explained as vulnerability, exposure, sensitivity, adaptive capacity, but also resilience. Commonly, adaptation to climate change refers to the engineering resilience concept. Similarly, when climate change adaptation examines a vulnerability of a species or habitat, the ecological resilience approach seems most appropriate and can provide a framework for identifying critical thresholds (Fünfgeld and McEvoy 2012).

Traditionally, the goal of risk mitigation is the protection of people, of properties and of the environment from the destructive force of catastrophic events (see Sect. 2.2). Nevertheless, building a resilient city does not only concern the use of land and structural engineering, but must also provide the capability to anticipate and respond to disasters. Static engineering-based conceptualisation of resilience has been embraced for a long time by disaster studies and the focus has been on recovery in quantitative terms. On the contrary, the metaphor of social-ecological resilience helps to synthesise the integration between ecology and social sciences due to the central role of the human ecosystem framework in multiple scale.

Going back to planning, we could identify three characteristics that allow the use of the social-ecological resilience concept for urban and territorial systems:

- humans are part of the ecosystems.

> The models derived from the human ecosystem framework are not a matter of humans versus nature, but humans and ecological processes combined into a reciprocally interactive network (Pickett et al. 2004, p. 378)

- territorial systems are complex systems. They are the sum of several interacting subsystems.
- a territorial system is characterised by adaptive cycles. The ecological, social and economic processes permit the continued adjustment and self-organisation of urban systems.

Thus, in the planning process, resilience highlights the need to be more flexible. It considers transformation as normal, and dynamism is an intrinsic factor of how systems act. For this reason, a deep affinity appears with governance theory and social learning, co-management and participation.

As Godschalk (2003) argued

> a resilient city is a sustainable network of physical systems and human communities (ibid., p. 137).

In this metaphor, the physical systems are the constructed and natural environmental components of the city, whereas human communities concern all the actors that live, work and act in that space. The physical systems include structural engineering like road networks, energy facilities or infrastructure and building, as well as natural systems as topography, geology and soils. For Godschalk, the physical system is the body of the city, its bones, arteries, and muscles. As a result, the physical systems must work in order to guarantee a proper and effective functioning under extreme stress during a disaster, otherwise a city would be extremely vulnerable to disasters without a persistent and resilient physical system.

In the same way, a body without a brain directing its activities, responding to its needs, and learning from its experience is likewise fragile and vulnerable. The metaphor of the brain refers to the human communities with the social and institutional components of the city.

> They include the formal and informal, stable and ad hoc human associations, that operate in an urban area: schools, neighbourhoods, agencies, organisations, enterprises, task forces, and the like. (…) During a disaster, the community networks must be able to survive and function under extreme and unique conditions (ibid., p. 137).

Social and institutional networks have varying degrees of organisation, identity and cohesion. During a disaster they must be able to handle emergency. Their absence or inefficiency undermines the capability to find solutions.

If we take on Beck's notion of reflexivity in risk society it could be very close to the meaning of resilience (Sapontzaki 2007, see Sect. 2.4). Despite the growing interest among the politicians, and practitioners to use community resilience as a means of confronting the response to local disasters (McAslan 2010), resilience remains a difficult concept to practice, thus, achieving a tangible outcome is challenging compared, for instance, with the risk management concept (Mitchell and Harris 2012). Nowadays most of the actions in this context are focused only on making the system physically resistant to the force of disaster (Godschalk 2003), but there are not enough programs for community preparedness, forecasting and warning (Pearce 2003). If community resilience is improved, it would be more responsible for building ultimate urban resilience.

The key challenge should be built in knowledge, learning ability and adaptation of institutions that manage ecosystems at different levels, to improve the resilience of natural and human systems and to contrast their vulnerability in the present context of uncertainty.

The adaptive cycle—as defined by Holling– concerns the characteristics of being a systemic process of complex environmental systems, characterised by high levels of uncertainty, and affected by potential ecological social and economic impacts, due to different management options. This process must continually improve its management skills, to constantly learn and adapt.[2] The ability to learn is related to the cyclical approach that manages the effects of policies and/or actions and includes the results in subsequent decisions with the integration of different knowledge. One useful outcome of the resilience method is exactly the ability to link together phenomena that in mainstream planning approaches remain firmly separated.

In this view, an alternative theoretical perspective of a more dynamic, fluid and interpretive planning that requires a shift of methodology from *command-and-control* to *learn-and-adapt* is drawing (Davoudi et al. 2013). For this purpose, planning as learning is fundamental to align ideas and goals, to shape leaderships and achieve a shared vision instead of acting as a bureaucratic process (Teigão dos Santos and

---

[2]Pickett et al. (2004) use the concept *learning loop* to define the need of long-term dialogue among different institutions and community in order to ensure the monitoring and implementation of knowledge.

Partidário 2011).[3] Stakeholders must create new relationships to enhance multidirectional information flows to learn from each other and to develop together flexible ways of managing their environments. In the same way, planning also as communication is crucial to enhance information flows and to build relationships. Planning should be a process of communication with its regulations, orientations and actions, but in reality, the information displayed at the end of the planning process often disregards the different capacities and needs of the potential end-users. Therefore, planning for resilience could be considered as more perspective to foster adaptive capacities. The different stakeholders as well as the citizens and the decision-makers are central in this process.

In this research, social-ecological resilience is defined, hence, as the social capacity to respond to disturbances and changes. The analysis is further oriented to organisational and institutional dynamics underling ecosystem management.

3.5 Linking Vulnerability and Resilience

Though related, vulnerability and resilience are different concepts in the field of the contemporary challenges posed by environmental changes, social context or global economy. Usually, the terms vulnerability and resilience refer to two different approaches to the problem world of today: the former is more related to the ecological-biophysical dimension, whereas the latter focuses more on the social political issues (see Sect. 2.3). As Miller et al. (2010) argued, the epistemological tension between the two terms is mostly caused by the research communities related to these terms and how either communities conceptualise systems and changes. Miller also theorised that vulnerability and resilience are complementary, exploring the need of integration between the two terms to reach a set of common conceptual and methodological principles able to guide new claims for future management and governance.

The term vulnerability has been theorised in hazard studies and in response capacity in regional sciences (Eakin and Luers 2006; Smith 2009; Simmie and Martin 2010). On the other hand, the epistemic and academic contributions to resilience theory primarily come from natural sciences (Holling 1973; Pendall et al. 2010; Simmie and Martin 2010), but nowadays it is spread also to social sciences (Berkes and Folke 1998; Folke 2006; Gunderson and Holling 2002). Vulnerability is seen as a

[3] Teigão dos Santos and Partidário identified four different characteristics that planning processes should adopt to be resilient: anticipation, innovation, learning and communication.

(a) Planning as anticipation is important to perceive emergent disturbances and to identify early signals and possible solutions.
(b) Planning as innovation is needed to be more flexible and dynamic in generating better solutions, instead of repetitive processes.
(c) Planning as learning is fundamental to knowledge, to produce consensus and co-accountability.
(d) Planning as communication is decisive to raise consciousness, to bring support and build relationships.

condition related to exposure, susceptibility, sensitivity, coping and adaptive capacity related to *calculated risks*. Therefore, differently by resilience, vulnerability is shaped by dynamic historical processes, differential entitlements rather than being a direct outcome of a perturbation or stress (Eakin and Luers 2006; Miller et al. 2010). Resilience, instead, includes different critical features as persistence, adaptability and transformability to *uncalculated risks* (Resilience Alliance 2010). The core of resilience lies in the interactions among system components and consequently in the institution role, in social capital, in leadership and learning. In risk management, the resilience approach has increased, and it is always more applied to institutional responses to climate change solutions in livelihood transformations.

To summarise, there is a common ground between vulnerability and resilience that concerns the way systems respond to stress or perturbations, while the difference can be found in the way each community considers systems. Vulnerability focuses on the understanding of ecological and biophysical processes—in particular the distribution of costs, risks and benefits—in order to anticipate, adapt to and manage change. Resilience instead emphasises system dynamics and interconnections, social-ecological relations and feedback.

However, the many definitions and approaches to vulnerability and resilience reveal the multifaceted nature of the problems under examination. Because of this multidisciplinary and fragmented nature of the topic and the absence of a dominant paradigm within the literature, the following chapter is not meant to be exhaustive, but aims to define the theoretical framework of this research.

In particular, this research suggests maintaining the essential role of planning as decision-making activity in managing the future of a society. Nonetheless, the shifting paradigm in risk perspective requires an alternative path of planning to maintain its role; a path that acknowledges the implications of social security where the assumption is not to control a defined and measurable risk, but to make an uncertain future manageable.

3.6 Criticism to Resilience

Just like its success, the resilience concept is also the subject of widespread criticism. Its elusive character has given the opportunity to widely use it in different disciplines with varying degrees of success, but at the same time its attractiveness risks its incorrect usage for scientific justification of particular policies, projects and practices (Olsson et al. 2015). There are several criticisms to resilience including its excessive positive connotation, the potential implicit devolution of responsibility and the implications for more vulnerable and poor communities.

Firstly, there is a tendency to understand resilience as positive (Meerow et al. 2016). Both ecological and socio-ecological currents often suggest a positive meaning that implying that resilience is desirable and *good* (Chandler and Coaffee 2017). But resilience has both dark and bright sides; it contains some contradictory and even negative aspects because resilience reveals itself only after failures of development and policies. Similarly, if we look to the semantics of the resilience concept, being resilient might also mean being an obstacle to positive change, however, it is often employed to describe the capacity to react to disaster, terrorist attacks, market shocks or sexual abuse. The positive connotation turns these policy failures into learning opportunities.

Another oversimplification is the tendency to see resilience in a binary way as a concept that there is or there is not, without any scaling understanding (Bourbeau 2013). This contradicts the complexity of the world to which resilience applies. For example, resilience might be dependent also on time and type of disturbances; a society may be able to deal with certain adversities but not with others.

However, the most famous critique on resilience refers to the relationship between resilience subjects and neoliberal forms. This especially characterised the literature of political science and international relations that look at resilience as a form of governmentality. This paradigm suggests the potential close connection between the rise of resilience and the dominance of the neoliberal approach, according to the Foucauldian-inspired interpretation. In modern neoliberalism, there is a shift from the state as provider to the state as enabler and promoter of self-reliance (Chandler 2014). The government encourages the civil society to be free to take the initiative and responsibility in the areas of risk and security. Self-awareness, reflexivity and learning of resilience thinking are considered ways to encourage the idea of active citizenship in a liberal form of governance. Furthermore, neoliberalism scholars argue that resilience expresses a change in the political strategy: from a centralised prevention to an individual or a community scale (Reghezza-Zitt et al. 2012). It is also being argued that international agencies have contributed to promoting the need of responsibility of local governments as well as individuals, with a radical change in the approach to disaster management.

Finally, the shift of scale in disaster management introduces a third critique. In resilience literature, the adaptive capacity of social systems seems to be dependent on the nature of their institutional entities and their capabilities to deal with shocks; this can be considered as an intrinsic property of the system that may differ from context to context or it can refer to how well a society is navigating through some past adversity. This shift of scale that puts responsibility mainly on the local stakeholders, penalises those communities that are more vulnerable and are not able to organise themselves or to take action.

References

Adger WN (2000) Social and ecological resilience: are they related? Prog Hum Geogr 24:347–364

Berkes F, Folke C (1998) Linking social and ecological systems for resilience and sustainability. Link Soc Ecol Syst Manag Pract Soc Mech Build Resil 1–25

Berkes F, Colding J, Folke C (2003) Navigating social-ecological systems: building resilience for complexity and change. Cambridge University Press, Cambridge

Bertuglia CS, Staricco L (2000) Complessità, autoorganizzazione, città. Franco Angeli, Milano

Bourbeau P (2013) Resiliencism: premises and promises in securitisation research. Resilience 1:3–17
Brundtland G, Khalid M (1987) UN Brundtland commission report "Our Future"
Chandler D (2014) Beyond neoliberalism: resilience, the new art of governing complexity. Resilience 2:47–63
Chandler D, Coaffee J (2017) The Routledge handbook of international resilience. Routledge, New York
Coaffee J, Wood DM, Rogers P (2009) The everyday resilience of the city: how cities respond to terrorism and disaster. Palgrave Macmillan, Basingstoke [England]; New York
Crutzen P (2002) Geology of mankind. Nature 415:23. https://doi.org/10.1038/415023a
Davoudi S, Shaw K, Haider LJ et al (2012) Resilience: a bridging concept or a dead end? "Reframing" resilience: challenges for planning theory and practice interacting traps: resilience assessment of a pasture management system in Northern Afghanistan urban resilience: what does it mean in planning practice? Resilience as a useful concept for climate change adaptation? The politics of resilience for planning: a cautionary note. Plan Theory Pract 13:299–333
Davoudi S, Brooks E, Mehmood A (2013) Evolutionary resilience and strategies for climate adaptation. Plan Pract Res 28:307–322
Eakin H, Luers AL (2006) Assessing the vulnerability of social-environmental systems. Annu Rev Env Resour 31:365–394
Edson MC (2010) Group development: a complex adaptive systems perspective. In: Proceedings of the 54th annual meeting of the ISSS-2010, Waterloo, Canada
European Commission (2005) Frontier research: the European challenge. High level expert group report Brussel
Folke C (2006) Resilience: the emergence of a perspective for social–ecological systems analyses. Glob Environ Change 16:253–267
Folke C, Carpenter SR, Walker B et al (2010) Resilience thinking: integrating resilience, adaptability and transformability. Ecol Soc 15:20
Folke C, Hahn T, Olsson P, Norberg J (2005) Adaptive governance of social-ecological systems. Annu Rev Environ Resour 30:441–473
Fünfgeld H, McEvoy D (2012) Resilience as a useful concept for climate change adaptation? Plan Theory Pract 13:324–328. https://doi.org/10.1080/14649357.2012.677124
Glaser M, Krause G, Ratter B, Welp M (2008) Human/nature interaction in the anthropocene. GAIA 17(1):77–78
Godschalk DR (2003) Urban hazard mitigation: creating resilient cities. Nat Hazards Rev 4:136–143
Gotts NM (2007) Resilience, panarchy, and world-systems analysis. Ecol Soc 12:24
Gunderson LH, Holling CS (2002) Panarchy: understanding transformations in systems of humans and nature. Isl Wash
Gunderson LH, Holling CS, Light SS (1995) Barriers and bridges to the renewal of ecosystems and institutions. Columbia University Press
Gunderson LH, Allen CR, Holling CS (2009) Foundations of ecological resilience. Island Press
Holling CS (1973) Resilience and stability of ecological systems. Annu Rev Ecol Syst 1–23
Holling CS (1996) Engineering resilience versus ecological resilience. Engineering within ecological constraints. National Academy Press, Washington, DC, USA, pp 31–34
Holling CS (2004) From complex regions to complex worlds. Ecol Soc 9
Jahn T, Becker E, Keil F, Schramm E (2009) Understandind social- ecological systems: frontier research for sustainable development. Implication for European Research Policy
Lambin EF (2005) Conditions for sustainability of human–environment systems: Information, motivation, and capacity. Glob Environ Change 15:177–180
McAslan A (2010) Community resilience. Understanding the concept and its application. Torrens Resilience Institute, Australia
Meerow S, Newell JP, Stults M (2016) Defining urban resilience: a review. Landsc Urban Plan 147:38–49

Miller F, Osbahr H, Boyd E et al (2010) Resilience and vulnerability: complementary or conflicting concepts. Ecol Soc 15:11
Mitchell T, Harris K (2012) Resilience: a risk management approach. ODI Backgr Note January Lond Overseas Dev Inst
Olsson L, Jerneck A, Thoren H et al (2015) Why resilience is unappealing to social science: theoretical and empirical investigations of the scientific use of resilience. Sci Adv 1:e1400217. https://doi.org/10.1126/sciadv.1400217
Ostrom E (1990) Governing the commons: the evolution of institutions for collective action. Cambridge University Press, Cambridge
Pearce L (2003) Disaster management and community planning, and public participation: how to achieve sustainable hazard mitigation. Nat Hazards 28:211–228. https://doi.org/10.1023/A:1022917721797
Pendall R, Foster KA, Cowell M (2010) Resilience and regions: building understanding of the metaphor. Camb J Reg Econ Soc 3:71–84
Pickett STA, Cadenasso ML, Grove JM (2004) Resilient cities: meaning, models, and metaphor for integrating the ecological, socio-economic, and planning realms. Landsc Urban Plan 69:369–384
Porter L, Davoudi S (2012) The politics of resilience for planning: a cautionary note. Plan Theory Pract 13:329–333. https://doi.org/10.1080/14649357.2012.677124
Reghezza-Zitt M, Rufat S, Djament-Tran G et al (2012) What resilience is not: uses and abuses. Cybergeo Eur J Geogr
Resilience Alliance (2010) Assessing and managing resilience in social-ecological systems: a practitioner's workbook, Version 2.0
Shaw K (2012) "Reframing" resilience: challenges for planning theory and practice. Plan Theory Pract 13:308–312
Simmie J, Martin R (2010) The economic resilience of regions: towards an evolutionary approach. Camb J Reg Econ Soc 3:27–43
Smith K (2009) environmental hazards: assessing risk and reducing disaster, 5 edn. Routledge, New York
Teigão dos Santos F, Partidário MR (2011) SPARK: strategic planning approach for resilience keeping. Eur Plan Stud 19:1517–1536
Walker B, Holling CS, Carpenter SR, Kinzig A (2004) Resilience, adaptability and transformability in social–ecological systems. Ecol Soc 9:5
Welsh M (2014) Resilience and responsibility: governing uncertainty in a complex world. Geogr J 180:15–26. https://doi.org/10.1111/geoj.12012
Westley F (1995) Governing design: the management of social systems and ecosystems management. In: Barriers and bridges to the renewal of ecosystems and institutions
Wilkinson C (2012) Urban resilience: what does it mean in planning practice? Plan Theory Pract 13:319–324. https://doi.org/10.1080/14649357.2012.677124

Chapter 4
Responding to Disasters

Abstract The chapter critically examines the contemporary structure of risk management by drawing on resilience perspective. Usually, disaster risk studies and emergency management are rooted in the PPRR chain (prevision, prevention, response and recovery) in which each step includes several actions, and competences are laid down by law. Its emphasis is on identifying *exposure* and *vulnerability* to risk and on bouncing back to normalcy. The result is a rigid process in which there is great attention given to the more predictable types of events and impacts, while less consideration is given to the interactions among actors across multiple scales and time frames. Network and cooperative relationships are taken for granted, even if they include a large number of actors with different skills. The central argument of the chapter is the key role that knowledge production, implementation and dissemination and the learning processes have in building resilience. The first part of the chapter analyses the multiple levels and sectors of interventions of the PPRR chain, illustrating the cross-scale challenges; while the second part problematises the integration of diverse disciplines knowledge and the role of collaborative knowledge production in a resilience approach.

Keywords PPRR chain · Emergency management · Competences · Knowledge systems · Collaborative knowledge production · Social learning

4.1 Resilience and Risk Management

This chapter aims to develop a conceptual framework that helps to re-think risk management by drawing on the three distinct perspectives on resilience (engineering, ecological, social-ecological or evolutionary), which are discussed in Chap. 3.

Usually, disaster studies largely embrace the engineering-based conceptualisation of resilience that is dominated by an emphasis on bounce back to normalcy. This approach works to make the system robust through an already stable physical infrastructure and pre-defined allocation of responsibility. It seeks resilience largely in the form of persistence, while social dimension seems to have a limited influence in enhancing or diminishing resilience.

E. Pede, *Planning for Resilience*, SpringerBriefs in Geography,
https://doi.org/10.1007/978-3-030-17262-6_4

This is in contrast with the concepts of flexibility, adaptability and transformability that are promoted by social-ecological and evolutionary resilience (Davoudi et al. 2013). From this perspective, human and nature are interdependent systems (Folke et al. 2010), and people can intervene through their technologies and innovations to adapt and transform towards a more desirable trajectory. In particular, the flexible use of resources, including social capital, and the cooperative network at different scales are considered main levers for building resilience.

This does not imply that persistence is not important. It remains essential for managing risks, particularly during and shortly after a disaster, but is not exhaustive, especially in today's world of uncertainty. For this reason, the social-ecological resilience framework overlaps the human component to persistence. We understand that humans operate directly or indirectly, consciously or unconsciously creating unexpected outcomes, options and opportunities, thus increasing their capacity to become more or less resilient.

The first section of the chapter explores the Prevision-Prevention-Response-Recovery chain to present a critique of the current approach to risk management highlighting its limitations by drawing on the four-dimensional framework for building evolutionary resilience as suggested by Davoudi et al. (2013).

From this perspective, resilience should be a dynamic interplay among persistence, adaptability, transformability and preparedness across multiple scales and time frames. Persistence is the main component of engineering resilience and it refers to the robustness of a system. Adaptability is drawn from ecological resilience and indicates the ability to adapt to new circumstances by enabling different resources. Finally, transformability and preparedness represent the quality of social-ecological and evolutionary resilience. The former implies the possibility to change path in development; the latter links the previous three: it harmonizes the different policies and practices outside the sectorial boundaries. This requires, not only technical and scientific knowledge, but also stakeholders and public engagement, including new actors and informal knowledge. The chapter problematises the importance of the institutional governance across scales and time frames, the role of knowledge and *learn-and-adapt* capacity in enhancing resilience. Its aim is to reference resilience, cross-scale and risk management literature, while also including additional knowledge management literature that is not commonly found in public policies.

4.2 The Prevision-Prevention-Response-Recovery Chain

Extreme events can involve several temporal, spatial and jurisdictional scales and consequently an enormous number of actors with different values, levels of knowledge and practice (Adger 2006; McEvoy et al. 2013). There is a great variety in the way risk management is organised in different countries, however, the common view considers risk management as being the result of several consequential steps that cover prevision actions to recovery ones, working as a unique cycle (Smith 2009). When we talk about extreme events, we usually refer to three distinct times (*before*,

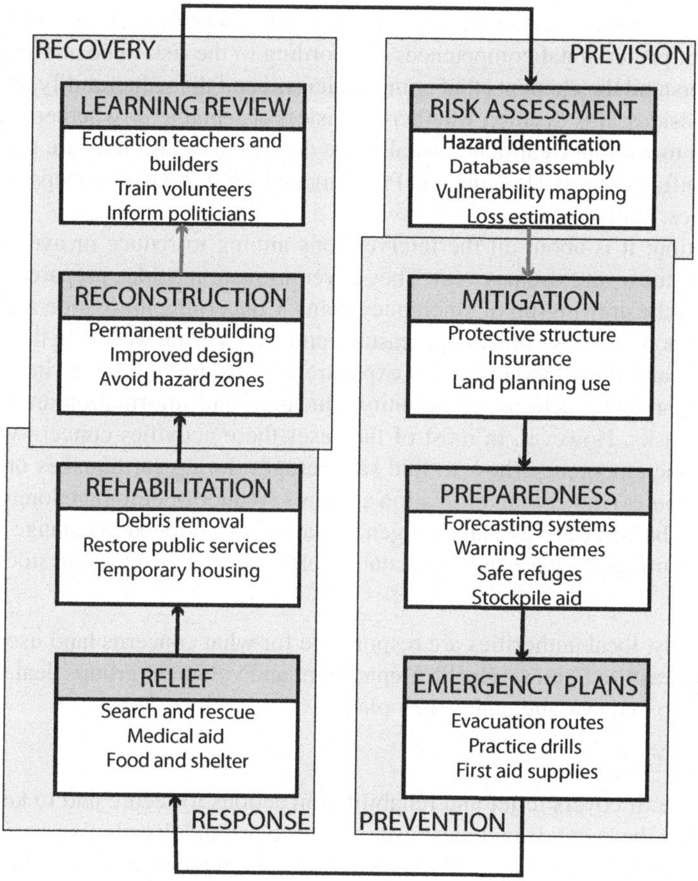

Fig. 4.1 PPRR—chain and its activities. *Source* Figure of the author first publish in Pede (2018)

during and *after* the event) with different phases in each. Firstly, there is the ordinary time, during which actors are involved to avoid or mitigate the possibility of an emergency situation, or however, to be prepared to deal with it. Then, when the event occurs the emergency phase starts. That is what we call the *golden* hours, referring to the ones immediately before and after and emergency relief operations. Finally, after a catastrophic event there is the time dedicated to the activities necessary to return to normality.

In order to categorise the treatments that can be deployed, Smith (2009) had subdivided each of these three phases into different steps that together constitute the PPRR (prevision—prevention- response—recovery) chain (Fig. 4.1):

Before the event

- Prevision: it includes the actions for evaluating and ranking the risks and for identifying areas that may be affected. It concerns the collection and processing of such data, determining the likelihood and consequence of each risk. In this phase, different professional competences—according to the risk to be analysed—work to understand the elements that compose hazard and the vulnerability of a specific area. Besides, it is decided whether the risk is acceptable or whether it is serious enough to warrant treatment. Usually, the competent authorities are the Environment, Infrastructure, Housing and Planning and Civil Protection Departments and Agencies, at diverse levels.

- Prevention: it is about all the interventions aiming to reduce or avoid potential damage due to an extreme event. They cover mitigation works, preparedness activities and the drawing-up of emergency plans. Concerning mitigation actions, they include both land-use planning measures and engineering works. In this stage, the actors deal with vulnerability and exposure modification of the territory. Regarding preparedness actions, community education and information are defined as crucial tasks. However, in most of the cases these activities concern *what to do* during the emergency (how to find safe refuges during earthquakes or floods or escape routes) rather than mitigation actions to reduce or eliminate long-term risk. Besides, hazard warning and emergency plans are elaborated to manage in the best way a contingent emergency (evacuation procedures as well as the stockpiling of supplies).

Generally, local authorities are responsible for what concerns land use and planning, whereas fire fighters, Health Department and volunteer groups deal with community preparedness and emergency plans.

During the event

- Response: it covers relief and rehabilitation actions to secure and to keep people safe after the event. It includes rescuing survivors and distributing basic supplies (food, water, medical care). The *response* depends on the magnitude of the disaster and possible cascade effects; it also relates to the flow of resources, equipment, information and knowledge that involved actors can deploy (Comfort et al. 2004).

Commonly, if the event is not considered critical, fire fighters are the main department involved; otherwise, civil protection, specific police forces (e.g. forest service in case of forest fire events), volunteer services or even military defence are involved. The coordination during this phase is not limited to the horizontal interaction among the forces involved, but it should also affect the competent authorities of previous steps of prevision and prevention.

After the event

- Recovery: it combines the first activities destined to re-enable the fundamental functions (e.g. removal of debris, reactivation of power and water supplies as well as infrastructure), to the reconstruction, where long-term actions are taken to support a return to normality after the devastation of the communities affected. This phase does not concern only physical activities but also includes psychological counselling.

If we examine the PPRR chain from a resilience perspective, its implicit view of resilience veers between the ecological and the more rigid engineering interpretation (see Sect. 3.1). The approach is dominated by an emphasis on bounce back to normalcy: prevision and prevention phases focus on risk assessment, vulnerability and exposure to the main risks. Most of the prevention actions are planned in terms of persistence to potential perturbations (through protective structure, safe refuges, or stockpile) or in preparation to overcome the events in the shortest possible time (forecasting systems and emergency plans).

The temporal order of interventions is characterised by rigid categorization of actions and linearity. On one hand, the pre-defined rigidity of competences helps to identify responsibilities especially during the short-term post-disaster responses; on the other, it can limit adaptability, resourcefulness and innovation.

However, it should be stressed that the PPRR chain is defined as a cycle, thus it foresees interventions aim to change the trajectory. Ideally, the recovery phase is defined as a moment to review the arranged risk management, to find out what has gone wrong and which measures need to be improved. It applies, to a certain extent, a *learn-and-adapt* methodology with an attempt to building a dynamic planning in which the different stages of the PPRR-chain are part of a continuous adaptive cycle. With respect to the social-ecological resilience concept, the reconstruction step could represent the "reorganisation" phase of the adaptive cycle in which changes and transformations occur in response to the unsustainable previous conditions that had co-determined the breakdown of the system.

> [...] effective risk resolution depends on the implementation of a sequential series of actions. The individual stages often overlap but it is crucial that they operate as a closed loop in order to draw benefits from experience and feedback (Smith 2009, p. 67).

In the practice, there is a limited attention to the adaptive dimension. One of the reasons could be rooted in the poor consideration to the social process and to the interconnections between the phases at multiple scales and time frames. The coordination is provided by the law, but it is underestimate. The emphasis is placed in the temporal order of interventions that must be consequential rather than their interconnection. Cooperation is mostly based on exchanges of information among institutions without an effective process of stakeholder engagement.

4.3 Disconnected Policies and Actors

As explained above, the PPRR chain involves a great many actors and covers a wide range of policies. As the phases, also the engaged responsibilities and competences overlap. Consequently, resolving risk problems requires dialogue and coordination among experts to allow stages to operate as a closed loop.[1]

---

[1]Also European Communities (2009) put *linking the different actors throughout the disaster cycle* as a key issue in the Commission's point of view. The environmental issue is maybe the first one that integrates the multidisciplinary approach as the only way to find long-lasting solutions.

Nowadays, even if there is progressive awareness of the need of collaboration among the actors and stakeholders of all phases, a fully integrated approach to disaster reduction is rarely achieved. There is a wide range of aspects that should be certainly considered in this regard: the number of actors, the complexity of risks, institutional weakness or fragmentation as well as the lack of technical expertise.

Generally, the risk management framework is set by government regulations operating at local, regional, national and even international levels. If we consider the EU context, most of the activities are usually fulfilled by civil protection. At the European level, the EU Civil Protection legislation was revised at the end of 2013 to increase the security of citizens through pre-planned and effective manners to respond to natural and man-made disasters. The new legislation places more emphasis on disaster prevention, risk management and disaster preparedness. It asks the Member States to share a summary of their risk assessments and to refine their risk management planning as well as to share the best practices (ECHO 2014).

Generally, the primary responsibility for dealing with a disaster lies with the country in which it occurs. But when the scale of emergency overwhelms national response capabilities, the EU Civil Protection Mechanism enables a coordinated assistance from the participating States. Each country has its own legislation on risk management and civil protection, however, most of the countries are based on the principle of vertical subsidiarity within the institutional level of power. At the same time, there is also horizontal interaction within the public and the private sector as well as the inclusion of volunteer organisations. At the national level, the relevant activity is usually quartered in a formal governmental agency mostly with policy and programmatic directive responsibilities. At other levels, the civil protection arrangements are very heterogeneous, mostly complex with problems of integration between higher and lower levels. However, the lower governmental levels are the most important actors,[2] with the support of external civil protection groups composed mainly by semi-volunteers.

Risk mismanagement can be affected by missing or failed institutions, or by scale mismatches among institutions (Costanza et al. 1993). These can be related to ineffective connections between scales or because decisions are based on information aggregated at the wrong scale. It should be considered that decision-making on risk is made more difficult by the very nature of risk: risks are combinations of complexity, uncertainty and ambiguity (Renn 2015). Complexity is given by external factors that can affect the casual relationship between cause and effect. Equally, uncertainty refers to the levels of confidence in available knowledge that stakeholders can have at their disposal in the process of decision-making. In other words, stakeholders act between a lower and an upper limit of what can reasonably be expected. Finally, ambiguity is related to the interpretative meaning of knowledge that each actor has, due to his background. This is one of the hardest obstacles to overcome. The gaps in

---

[2] In most of the countries there is an evolution to a de-centralized civil protection system (especially in Europe where the subsidiarity has had great influence), but there are still very centralized national level systems such as in Japan.

understanding exist between most professional areas, but can be exemplified by the following relationships (Lindberg and Sundelius 2012):

- Between security and safety professionals;
- Between different civil authorities;
- Between public authorities and the private sector;
- Between levels of authorities;
- Among individuals.

All these groups have the necessity to cooperate and coordinate, before, during and after a catastrophic event in common. Some key factors contributing to the misunderstanding and to the lack of or inefficient cooperation are rooted in the different terminologies, ways of organizing and finally in the procurements of sector-specific technological solutions that each category has. Not surprisingly, they have problems in understanding each other and the way their sectors/competences interact.

As Folke (2006) argued, the biggest change in this context must be built in the knowledge and learning capacities of institutions and organizations. In this respect, effective communication is the prerequisite for efficient and fair risk governance because it provides the knowledge base on which dialogue among stakeholders can be productive (Comfort et al. 2004; Renn 2015). Likewise, resilience thinking considers knowledge and learning capacity key elements in the process of continuously improving management skills and adaptation (the adaptive cycle, see Sect. 3.3). The role played by humans and their interaction is what makes the process more dynamic and holistic.

4.4 Responsibilities, Institutions and New Actors

As set out in Sect. 3.2, following resilience theory, we know that social-ecological systems (SESs) are open systems where all cultural, political, social, ecological, technological and other components interact at multiple levels (Resilience Alliance 2010). The two subsystems (ecological and social) can be seen as being composed of a spectrum of interconnected exogenous, slow, and fast[3] variables with a direct impact on the people (Chapin et al. 2006). According to Anderies et al. (2004), several entities are involved in social-ecological system interaction: the resources (territory), the resource users (citizens), the public infrastructure providers (institutions), the public infrastructure (engineering works and social capital), the institutional rules and the external environment (weather, economy, political system). The evolution of system elements has different speeds, but the system approach is holistic because it does not focus on a detailed understanding of parts, but on how key components contribute to the dynamics of the whole system. Through these interactions, social-ecological systems can self-organise, bringing out new configurations and making

[3]Slow and fast variables are strongly linked. The former refers to environmental capital and government, whereas the latter concerns ecosystem users. Usually, the dynamics of fast variables, such as crop production, are shaped by slow variables, such as soil organic matter (Walker et al. 2012).

possible adaptations. Hence, SESs are suitable for resilience assessment due to their capacity to self-recover or self-reorganise in response to disturbances.

People respond to social, environmental, and ecological changes of the system through a complex network of institutions,[4] which in turn affect slow and fast variables of the system. Institutions involve multiple sectors and scales and the interactions among these actors; in the same way, organizations and rules determine how people make decisions, and exercise responsibility.

In a resilient approach the main aim of institutions should be the protection and the well-being of the community, which is achieved by controlling the use of land and organizing and directing social behaviour. Understanding the behaviour and interactions of institutions, therefore, is crucial for understanding social-ecological interactions in SES and for reorganizing in favour of a more sustainable development (see Sect. 3.3).

The institutional configurations of the systems concern several and different types of actors and tools and depend on the structure of the social system governing and managing the SES. Following Chapin et al. (2006), it is possible to recognize different categories of institutions that directly or indirectly affect ecosystems:

i. Resource-harvest institutions: include the sectorial actors that are focused mainly on the management of goods (e.g. agriculture, forest, water, etc.). They usually include regulatory tools;
ii. Resource-conservation institutions: which administer choices to conserve and protect the ecosystem. The focus is on long-term conditions.
iii. Hazard-reduction institutions: which govern choices for reducing the societal impacts of natural hazards.
iv. Ecological externality-producing institutions: refer to heterogeneous social economic sectors pursuing development goals. These institutions have indirect effects on the ecosystem (industrial activities, construction of infrastructure).

Management and governance process among different institutions can actually be difficult to achieve because institutions have different perspectives in recognizing and solving problems (Schön 1983). Thus, to address contemporary environmental problems, it is important to understand how barriers to collaboration can be overcome.

Public management of risk is mostly a multi-discipline and multi-sector policy field characterized by a wide range of actors from civil protection to sectorial and spatial planning. The coordination is not only an inter-sector issue but it involves different jurisdictional, institutional and temporal scales among the different stages of the PPRR chain (Wilbanks 2006; Bignami 2010; Sapountzaki et al. 2011). The number and type of actors involved concern the type of disaster (e.g. forest fire, earthquake, floods, landslide, etc), the spatial extension and the type of damages on

[4]In resilience and SES literature, the term *institutions* refers both formal and informal statements. Formal institutions are codified rules such as constitutions, law, and property rights. On the other hand, informal institutions include social and behavioural norms typical of a specific community. This chapter is mainly focused on formal institutions and the way they cooperate together. Informal institutions will be explored later, with focus on the possibility to promote collaborative behaviour in risk situations including citizens during emergencies.

people and goods. In most countries, risk management is entrusted to a variety of institutions, which operate on different spatial scales, from national to local level, covering the different stages of the PPRR chain. Some actors work during all stages of the PPRR, others are involved only in few of them. In the same way, some institutions are directly involved in the crisis (e.g. emergency management activities); others indirectly affect the pressures on the territory and consequently influence its vulnerability (e.g. territorial strategic development).

The proof of this complex situation, characterized by the overlap of responsibilities, is demonstrated by the general lack of a cohesive policy able to provide a suite of strategies that could be integrated. Despite the circularity of the activities provided for in the PPRR chain, reality shows us situations where the respective information, knowledge and policy actions run parallelly without any linkages, feedback, and mutual interaction.

It does not mean that all the PPRR-chain is characterized by malfunctions, but in most of the cases the ability in coordination is linked to the individuals' will rather than to procedures and tools. Therefore, the analysis of the risk governance can provide a clearer understanding of power dynamics and conflicts among stakeholders and their ability either to promote or to undermine the resilience of the system (Resilience Alliance 2010). Obviously, there is nothing such as a one optimal risk management structure for all, but adequate network characteristics can facilitate or impede processes such as information sharing and collaboration opportunities. In analysing the network, it is important to take into account the number of relations between actors, the degree of centrality of actors within the network, the existence of cohesive subgroups and their degree of isolation (Resilience Alliance 2010).

4.5 The Cross–Scale Challenge

In addition to stakeholders' coordination, the multiscale nature of the problems related to risk management demands attention. Generally, we are all aware that the causes, the dynamics and the consequences of extreme events can be measured at different scales, whereas it is less clear the importance of scales in the decision-making process. Usually, we use the term scale to refer to the spatial, temporal, quantitative, or analytical dimensions used by scientists to measure and study objects and processes (Gibson et al. 2000, p. 219)

Furthermore, when we use scales for analysing, their extent and resolution could also be explicit. The former indicates the size of the spatial, temporal, quantitative, or analytical dimensions of a scale; the latter refers to the precision used in measurement.

In connection to the concept of scales, there is the notion of *levels* that indicates the units of analysis that are located along a scale. Commonly, levels have a spatial feature (e.g. small, medium or large-sized phenomena) and can also be related to time (durations, frequencies or effects).

Closely related to spatial scales are jurisdictional scales defined as clearly bounded and organized political units (e.g. municipality, regions, provinces, nations, etc.). In the same way, jurisdictional characteristics are linked to institutional arrangements that also have fitting inner hierarchy (Cash et al. 2006). Following the classification of hierarchy defined by Gibson et al. (2000), there are two types of hierarchy: inclusive and constitutive where inclusive refers to levels with no cross-interaction and organization, whereas, constitutive hierarchy adverts to levels that can interact as in a complex system. In such systems, phenomena occurring at any one level are affected by mechanisms occurring both at the same level as by levels below and above.[5] Thus, it is fundamental to examine complex systems and their disturbances from a multilevel perspective because they depend on understanding the constraints at all levels.

Choosing the scale or the level for studying phenomena in such systems may cause different problems. Specifically, there is not an optimal scale or level, but at each scale/level there are different pattern identifications, problems, explanations of phenomena or questions to be asked. For instance, phenomena that appear at one level of resolution may be lost at lower or higher levels. It comes to light how the scaling problems can be related to issues of scale and/or level as well as to hierarchies and players involved. Also the debate on resilience highlights this issue: in ecological and socio- ecological resilience systems can have multiple stable states that implies that system, redefines its structure by changing the variables and processes that control behaviour.

Moreover, if most of the attention given to scale in studies of human-environment interactions has focused on spatial, temporal and jurisdictional issues, there are also some topics not conventionally framed as a scale problem, but in which *scale* can affect policy processes (Cash et al. 2006; Kok and Veldkamp 2011). Knowledge is one of them.

If describing and explaining phenomena depends on scale, it means that knowledge is also related to scale; thus we can talk of *knowledge as a scal*e (Cash et al. 2006) or of knowledge limited to scale. In other words, knowledge can be defined as *a system of progressive embedding* (Koraeus 2008) of data and information. It means that data is created at different scales and is linked and interpreted by different actors (at different levels) to support decisions. For instance, in planning, plans and actions are grouped into hierarchical set ranking in which data and models are combined at different spatial and temporal scales, or in which information is extrapolated between scales and levels.

Cash et al. illustrates how fuzzy and, at the same time, complex the definition of knowledge is in relation to the level of the analysis. It is difficult to identify the scales of knowledge, but it is undeniable that knowledge has to do with scale. By its nature,

> focus on a single geographic scale tends to emphasize processes operating at that scale, information collected at that scale, and parties influential at that scale – raising the possibility of misunderstanding cause and effect by missing the relevance of processes that operate at a different scale (Wilbanks and Kates 1999, p. 608).

[5]It was the Panarchy theory of Gunderson and Holling (2002) to add to hierarchy theory the concept that all levels are considered as influencing each-other in a top-down and bottom-up hierarchy.

For this reason, decision-makers should be sensitive to the knowledge of multiple scales; as well as, in building knowledge, there should be different types of interaction between the owner of the data or information and the last user.

Generally, planning theory focuses its attention on phenomena that encompass many scales and many levels; terms as multilevel, interdisciplinary and integrated approach are crucial but there are misunderstandings in the role of scale in the decision-making process (see Sect. 3.4). There is a particular lack of consideration in the cross-level or cross-scale interaction, in which *cross-scale* refers to interactions across different scales (e.g. between spatial domains and jurisdictions) that can result in nonlinear dynamics. Changes in cross-scale may arise from the consequences of interactions or be caused by other variables (Cash et al. 2006).

The cross-level and cross-scale interactions especially emerge from planning practice—considering climate change—posing formidable challenges for understanding and forecasting system dynamics. To be able to come to terms with cross-scale and cross-level dynamics, it is also crucial in governing ecological-social problems (Cash et al. 2006; Buizer et al. 2011). A hot spot of the social ecological system literature is the management of them, in particular referring to adaptive management. As Buizer et al. (2011) explained

> rather than focusing on discovering reality, attention is shifted to developing methods that could facilitate the analysis of complex SESs (ibid, p. 5).

Scale is not a new subject in planning; actually, it has always had a key role. The issue of scale and level interaction is usually considered in term of governance (Termeer et al. 2010; Kok and Veldkamp 2011): to identify and assess the influence and importance of key actors (stakeholder analysis) or to analyse the stakeholder relationship management. Over the years, several methods have been studied to understand stakeholders' behaviour (e.g. actor network analysis; social network analyses; system dynamic models) and all those works that acknowledge the idea that scales and levels must be considered as co-produced in processes in which institutions work together. Furthermore, in the literature and the practice related to scale and governance, there is a growing focus on the collaboration and integration among scales and levels that includes not only scientists, policy makers and shareholders but has an increasing interest in citizens and society.

In the governance debate, there seem to be two separate communities of study on the scale-related issue: the former mostly analyses the role of temporal and jurisdictional scales with a strong spatial perspective; whereas the latter focuses more on the role of networks and information flows released from spatial issue (Kok and Veldkamp 2011). Despite all commonalities, the proposed solutions differ considerably. The former approaches are more related to the polity dimension focused on the norms that allow the action of actors and that contribute to the multilevel governance effectiveness. The latter focuses more on communication claims and emphasize the way of policy making. However, most governance analyses focus only on a part of the policy processes. For instance, there is a lack of interest in exploring problems of horizontal and vertical interplay that concern building knowledge. In this case, the knowledge issue is not interpreted as the recognition and discussion of knowledge but more related to the sharing and interaction of data and information at different scales that are fundamental for comprehension (Buizer et al. 2011). More in detail, governance implies that many actors in different institutions contribute to policy development and implementation. Each of them has different knowledge claims but mutual interdependences that require cooperation and coordination among actors located at different levels. It means that knowledge produced at one level influences processes at another level, but it should be considered that actors have different knowledge systems due to their discipline background (Pahl-Wostl 2009). The domain of disciplines and their constituent parts affects the building knowledge processes. The plurality of views renders knowledge understanding and sharing complex and overflowing with obstacles. Consequently, these gaps in knowledge systems can compromise or create incongruences in cross-level interactions (Young 2006).

The implementation of integrated knowledge systems has a crucial role in effectively responding to the challenge of managing complex social-ecological systems. Therefore, the alignment and cooperation of capabilities to create, verify, absorb, share and, apply new knowledge can be one of the crucial issues to improve the resilience of complex systems to natural hazards (Roux et al. 2006).

4.6 Data, Information, Knowledge

Knowledge has always been implicated in all policy processes; decisions depend on the knowledge resources that are at the disposal of policy makers. The uncertainty and unpredictability of the society today stresses the limits of knowledge and consequently the limits of prediction in a world of complexity (Rothe 2017). Despite the unpredictability of the world, the primary task of planning is still addressing, overcoming or, at least, reducing uncertainty and in doing so it requires, more than ever, knowledge resources for developing effective responses to unexpected events. To examine how knowledge management can be applied, it is firstly crucial to explain what knowledge means.

Generally, a distinction is made between three/four main components: data, information, knowledge, and sometimes wisdom. Commonly, the terms data and information are incorrectly considered to be exchangeable with knowledge. As Koraeus (2008) argued

> A common trait in defining these concepts is to use a system of progressive embedding, where each step adds a layer of interpretation and contextualisation, but where each such addition also increases the level of cognitive complexity (ibid, p. 33).

Data can be considered as raw facts, the basic unit in the knowledge process. When data is organized in a given context, it creates information. In the same way, when information is interpreted, we obtain knowledge. Polanyi (1966, 2009) was the first to define knowledge as the process of interpreting and understanding data and information in order to take the right decision and to implement the right actions. Knowledge could be considered as data and information combined with context, interpretation and reflection. Additionally, it is possible to identify one more category: wisdom that is the utilisation of accumulated knowledge.

Data is a set of values of qualitative or quantitative variables. It refers to the measuring and the collection of numbers or characters, but it does not concern their analysis and interpretation. On the contrary, information refers to organized and interpreted data. The information process includes human participation but the outcome is explicit and can be readily transferred to another entity without any loss of meaning (Roux et al. 2006).

Finally, knowledge is a mix of experiences and contextual information that provides the basis for the decision-making processes. It means that knowledge is a cognitive process where users have a crucial role in determining the outcomes. But the human factor is crucial in building knowledge as well as in knowledge sharing. Knowledge is influenced not only by creators but also by users because they are the ones that must interpret the meaning.

As it highlights by social-ecological resilience, the human factor is not a second issue. Also in the field of computer system development, users are included in the information system. Langefors in *Essays on Infology* (Langefors 1996) introduced the infological equation to define process (i) where data (D) is a set of data that, depending on the pre-knowledge (S) a human experiences during her/his life and the time (t) available, is produced as information (I) (knowledge).

I i (D, S, t)
D the sets of data
S the user and her/his pre-knowledge
t the time available to the user for interpreting the data D
i the information conveyed by the data D
I the information function (knowledge)

(Langefors 1979, pp. 22)

The equation demonstrates that data sharing is successful only if users of the intended information system are included in the design process. If the users are neglected in the data definition phase there is a risk of data misinterpretation or wrong inferences could be drawn.

Also in relation to human agent, Polanyi classified knowledge into two types: explicit and tacit. The former is knowledge that can be captured in a database (*knowing what*). The latter is much more difficult to formalize, and it is linked to *knowing how*. More in detail, explicit knowledge refers to information and it can only partially represent what we know, because it cannot express the knowledge of experience. Contrary to explicit, tacit knowledge is much less *concrete* and is *non-codifiable* (Tsoukas 2005); it is rooted in practice but it interacts the norms of the disciplines and the knowledge for the practice. It is the knowledge that practitioners have but may not be able to describe and for this reason it is difficult to formalize and share with other people (Roux et al. 2006). Explicit and tacit knowledge are popular terms in management studies (Ambrosini and Bowman 2001; Nonaka 2008; Wenger 1998), but they are now spreading to geography and urban studies.

They both refer to different knowledge systems that are widespread in decision-making and governance processes. Generally, scientific (explicit) knowledge is considered the dominant lens through which decisions are taken. However, tacit knowledge is always present because it is related to stakeholders' skills. Usually, practitioners have to combine the knowledge resources at disposal with their intuitive understanding gathered from years of experience and practice. In the same way, the collective memory also should be considered as useful for the process of social learning. It concerns the knowledge hold by the community that usually is strongly rooted in a particular place and it can improve the scientific basis and fairness of community-based decisions.

4.7 Building and Sharing Knowledge

As mentioned above, in risk management diverse types and sources of knowledge are shared by diverse actors. This plurality of views can be problematic because there are differences in how people from different knowledge systems perceive and address conflicts and solutions (Zaucha et al. 2016). Identifying opportunities to bridge different knowledge systems can make important contributions towards understanding, governance and learning. In this respect, Renn (2015) recognized diverse typologies of knowledge that correspond to different types of *knowledge creators* (Roux et al. 2006):

- Systematic knowledge provided by fundamental or basic researchers: knowledge is systematized according to disciplinary lines. The main aim is the progress of understanding rather than the response to specific problems.
- Experiential knowledge that could come from:
  - applied researchers: knowledge creation is driven by practical problems. The knowledge product is shaped on the potential users' needs.
  - policy-makers: this knowledge requires inputs from different fields to support decisions for solving problems in practice.
  - operational managers: they use both explicit and tacit knowledge for organizational issues. The explicit knowledge comes from policy-makers' products; indeed, tacit knowledge is based on experiential learning and acquisition of good practices.
- Local and circumstantial knowledge provided by communities: this knowledge is based on traditions and experiences and it is transferred in social memory.

Risk problems require deliberation among knowledge holders able to produce a common understanding for facing indeterminacy, complexity and uncertainty of the real world. As Crosta argued

> building and processing the useful knowledge for policies can be divided between different workers but it cannot be separated (1998, p. 15 own translation).

This means knowledge in action is the result of different knowledge. Professional pluralism poses the predicament for the practitioners who must devise their own way of combing the different knowledge. Usually institution sectors and levels are interlinked by rules, procedures, organizational forms and technologies, which does not necessarily imply full interaction. For instance, in the field of risk management, it is not easy to describe the dynamic relationship among actors because each of them generate different types of demands that lead to the formation of as many types of response patterns based on diverse capacities (Comfort et al. 2004).

For example, if we consider planning practice, the plan process should be the field of interaction among actors but this does not necessarily imply an *interactive process* (Crosta 1998).[6] Planning is an interactive process when it combines shared procedures and norms with a common significance framework and with the practitioners' willingness to learn.

In his studies on *reflective practice*, Donal Schön analyses knowledge in practice, the nature of learning systems and the relevance of creating *learning society*. With the term *knowledge-in-practice*, Schön problematizes the complexity of knowledge useful for planning. In practice, planning knowledge involves the theory and the norms for planning practice, but it also has to deal with the objectives and goals of the practice, the values, the interests and the know-how by which planners contextualise the situations (Schon 1982). What stands out is the ever-changing and dynamic nature of knowledge according to the contexts.

In the same way, the changing world we are living in adds complexity to the practices and institutions must learn to become capable of learning, transforming and adapting themselves in response to changing situations and requirements. Consequently, institutions must become *learning systems*, systems capable of bringing about their continuing transformations.

> The loss of the stable state means that our society and all of its institutions are in continuous processes of transformation. (…) We must learn to understand, guide, influence and manage these transformations. We must make the capacity for undertaking them integral to ourselves and to our institutions. (…) We must, in other words, become adept at learning. We must become able not only to transform our institutions, in response to changing situations and requirements; we must invent and develop institutions which are 'learning systems', that is to say, systems capable of bringing about their own continuing transformation (Schön 1971, p. 28–29).

In Schön's learning systems, we can recognize the same significance of flexibility, feedback and organisational transformation that is assumed in the adaptive cycle of social-ecologic system resilience. Moreover, his work contributes to drawing attention on learning as a social action rather than individual, in which the government acquires new capacity by interacting among stakeholders building incrementalistic models of practices. Therefore, the challenge is how to share knowledge among stakeholders and the potential role of local communities as knowledge holders.

---

[6]The notion of interactive knowledge emerged not like a positivist paradigm but more like a strategic paradigm oriented to consensus building (Palermo and Ponzini 2014).

4.8 Knowledge Management

Usually, knowledge is created, stored and comprehended differently at different levels and without a cross-level framework. In the same way, the flow of knowledge among stakeholders is far from optimal; synergy among all parties is frequently poor, characterized by misunderstandings, conflicts and inefficiency. Misinterpretations characterize all the knowledge systems. Although, most of the attention focuses on the shift between scientists and managers, conflicts still persist also inside single scientific fields as well as among management sectors (Rogers 1998). The result is fragmented knowledge that is of little or no use to respond to complex problems.

The increasing complexity of problems to be faced has created a strong need for deliberate and systematic approaches to cultivating and sharing knowledge. Despite knowledge flows being the subject of many debates in planning, especially regarding governance, knowledge management as organizational innovation is not widely debated. The concept of *knowledge management* (KM), as a framework for creating, using and managing the knowledge, information of an organization, is, instead, widely used in the private sector. The present section suggests the use of knowledge management as a path to increase the learning capacities of stakeholders in a resilient perspective.

Knowledge management was theorized by Ikujiro Nonaka and Hirotaka Takeuchi in 1995 as an enhancement and extension of the existing theories of organizational learning. However, before it had been practiced for a long time, mostly in an informal manner rather that in a deliberate and systemic manner (Cong and Pandya 2003; Easterby-Smith and Lyles 2003). Knowledge management refers to a set of management activities aimed to maximise the process of knowledge creation and integration, including processes of sharing knowledge (Bowditch and Buono 2005). The main goal of knowledge management should be to optimize resources that already exist in an organization or institution so that its members will be able to seek out, utilize and enhance their activities and processes and increase their performance. Its added value is related to the possibility to reduce time, expense and error as well as to avoid overlapping in information creation. Following this premise, KM in institutions can be seen as the ability to use collective knowledge through a process of knowledge generation, sharing and exploitation.

The core of managing knowledge is concerned with deciding with whom to share, what is to be shared, how it is to be shared, and ultimately sharing and using it Cong and Pandya (2003).

The knowledge management process involves both explicit and tacit knowledge and improves the capabilities to create an interactive learning environment, where members are encouraged to readily transfer and share what they know, internalize it, and apply it to create new knowledge. As referred to above, it is false to believe that exchanging explicit knowledge among diverse units is easy because several problems can affect this process. But it is undeniable that the transfer of tacit knowledge is even harder because it requires a much higher intensity of interaction.

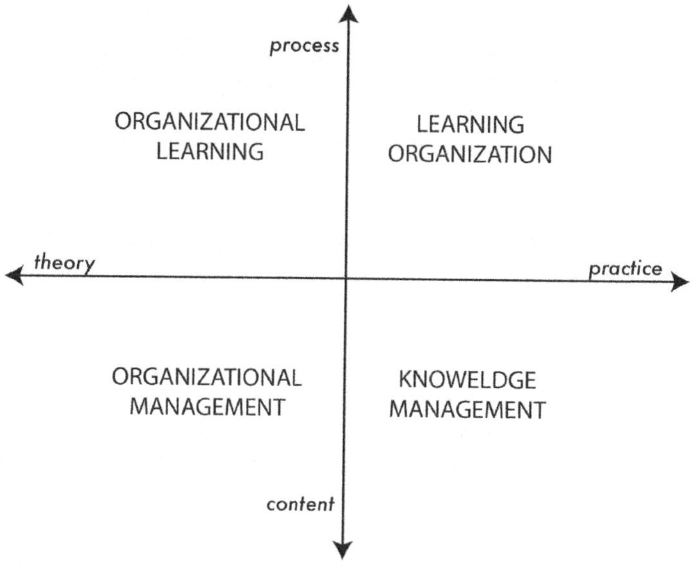

Fig. 4.2 Knowledge—knowing -learning matrix. *Source* Figure of the author

Knowledge sharing should consider four different concepts: organizational learning, organizational management, learning organization and knowledge management. They all refer to the process of creating, retaining and transferring knowledge but with different focuses and approaches. Easterby-Smith and Lyles (2003) summarized the distinctions among the four topics in relation to their attention on theory, process, content and practice issues (see Fig. 4.2). Organizational learning refers to the study of the learning processes of and within organizations, largely from an academic point of view, meanwhile, learning organization theory generally aims to understand how to create and improve this learning capacity. In the same way, the term organizational knowledge works for understanding and conceptualizing the nature of knowledge, whereas the knowledge management approach focuses on creating ways of disseminating and leveraging knowledge to improve the action.

In other words, KM is based on three key elements: people, processes and technology. People are the most important component, because managing knowledge depends on the individual's willingness to share knowledge with others. Generally, people tend to resist sharing their knowledge because creating knowledge is onerous in terms of time and money. Secondly, processes are the way to create, locate and share knowledge. Knowledge implementation should use systematic approaches so that organization can use what it knows to learn and create value. Finally, technology allows knowledge to be more accessible to people (Cong and Pandya 2003).

There are many benefits to be reaped from KM. In organizational settings, benefits can occur both at individual and organizational level. At the individual level, KM provides opportunities to enhance skills and experience by working together and sharing people's knowledge by learning from each other. This aspect is obviously crucial for the sharing of tacit knowledge, which is usually difficult to communicate and which requires extensive personal contact and trust to be effectively shared.

At the organizational level, KM provides two major benefits: it improves the organization's performance through increased efficiency, quality and reduction of time, and, at the same time, it enables better decision making thanks to the higher data integrity and greater collaboration between different sectors. Alongside this, the management of knowledge across government departments can also create new challenges for the retention of knowledge and the preservation of institutional memory.

Over the last decades, knowledge management has emerged as one of the most influential new organizational practices that has had wide diffusion especially in the private sector (Koraeus 2008). Despite the relevant literature for the private sector, there is a lack of analysis about KM in the public sector (Cong and Pandya 2003). There are several reasons accounting such low interest like the differences in human resource management and decision processes. Public institutions are traditionally compartmentalized structures in which the different departments comply with specific political priorities. The lack of knowledge management processes is the evidence of the inadequate integration among policies; and this attitude can create mismatching in the elaboration of the meaning of information and can hamper the sharing of knowledge across different units and levels.

As already mentioned, the obstacles to knowledge sharing concern how knowledge is generated, validated and stored. Contrary to traditional roles of knowledge provider and knowledge consumer that work through the unidirectional transfer of information, in sharing knowledge, knowledge should be "co-produced" through collaborative learning between experts and users creating unified learning systems.

> [...]this requires a shift from a view of knowledge as a "thing" that can be transferred, to one of a "process of relating" that involves careful negotiation of meaning among partners" (Roux et al. 2006, p. 1).

The processes and techniques for managing knowledge should follow different stages: from the identification of the knowledge domain and the gap between the existing and needed knowledge, to the selection of knowledge that seems appropriate, the consequent classification and storage of the filtered knowledge, and finally the review and the constant update to make it available to users. However, it involves not only the technical knowledge, but also requires political will and mobilization.

4.9 Role of Technology

As mentioned above, information technologies (IT) facilitate the creation, codification, transfer and integration of information. Before looking at the potential role of technology in more detail, it is important to underline such information technology can support KM but does not determine its success. Technology is a supporting innovation for people and processes; the three elements are interdependent parts of a whole and only the interaction between technology, techniques, and people, allow knowledge to be managed effectively. For instance, the conversion between data and information is efficiently handled through information technology, but IT is not able to turn information into knowledge. Information is interpreted into knowledge only through people.

Certainly, including technology has multiple benefits; first and foremost, the efficiency of data processing and data mining. IT can capture, store and distribute information quickly. However, information technology does not simply automate existing tasks but often helps to redefine them, even opening new possibilities (Kallinikos 2001). For example, building up a data infrastructure requires firstly a reflection on what kind of data is critical and should be collected and how. Then, data must be validated and evaluated.

Another benefit of using technology concerns standardization and control. The information produced by a computer-based information system is characterized by uniqueness, meaning it is expressed in a language of description that is intrinsic to such system and that should be common to all the actors who have access to it. To this aim information must be standardized and obey certain rules, otherwise, the systems might not be able to recognize or understand it.

However, the problem of interpretation still remains. Scientists interpret data based on the relevant experience and knowledge they have acquired; thus, different scientists can interpret the same data in different ways. Data interpretation involves constructing a logical scientific topic able to explain data, methods and techniques used to measure; otherwise, information can seem in contradiction if scientists do not share a common base line.

Over the last decade, the use of information technology in managing knowledge is constantly evolving. Before mass use of Internet, spatial data were stored in different physical locations, mostly using different standards or formats. Usually, one authority develops some spatial data content, while, all the other data that the authority needs comes from external sources. The amount of spatial data available and the usage of GIS have boosted more and more public and private organizations in sharing data both internally and with other organizations. In particular, the advent of the Internet and digitalisation have opened considerable new possibilities for users to access and utilize the data. In risk management, large interest has been focussed on the *Spatial Data Infrastructures* (SDIs): infrastructures, or basic physical and organizational structures, needed to facilitate efficient and effective use of spatial data for multiple users (Hendriks et al. 2012). SDIs work with spatial data from many sources to support a broad range of applications; SDIs potentially deliver products or support communication and cooperation processes. In other words, an SDI interconnects multiple GISs via web integrating their respective spatial data in a collaborative framework.

To interconnect several GISs, spatial information and relationships must be encoded in structured formal representations. This formal encoding makes these infrastructures inherently well suited to facilitate collaboration among technicians and experts from different fields but with common aims. Data harmonization and integration basically face two types of heterogeneity: data heterogeneity and semantic heterogeneity. The former refers to differences of data in terms of data type and formats, whereas, the latter refers to the meaning of the data according to the specific context. The transfer of data from one system to another does not guarantee that data has a meaning to the new user; it requires the sharing of meaning between the two users and furthermore that the two sets of meanings are identical.

Nowadays, geographic information systems are adopted widely but often independently. As a result, it is very difficult for different systems to share data. The effective integration of data systems is one of the key challenges of governing dynamic and socioecological systems. This requires more than just institutional relationship; it should be based on strong relationship-building and participatory process of stakeholder engagement able to activate process of social learning.

4.10 Social Learning

In the literature on resilience, the increase of uncertainties, the rapid dynamics of socioeconomic development and globalization have often been discussed as needing more adaptive and flexible policies able to integrate the capacities of many stakeholders. The mutual dependence between groups of experts has modified the government's notion, in favour of multi-scale and polycentric governance approaches where different stakeholders collaborate in the formulation and implementation of public policy (Pahl-Wostl et al. 2007b). Several authors have embraced the importance of the human dimension and the necessity of an adaptive form of governance that is operationalised through adaptive co-management emphasizing the role of networks, leadership, diversity, collective memory and trust. A central concept in such adaptive management is social learning (Folke 2006).

The social learning theory emphasizes both issues of knowing and issues of being part of and becoming. Originally, social learning was studied in psychology and sociology (Bandura 1977), focusing on how we learn. Later, it became a promising frame for understanding governance (Folke et al. 2002; Berkes et al. 2003). At its first conceptualization, social learning referred to the individual process of learning that takes place in a social context, but it did not consider group processes according to the consideration that learning process starts with the living experience and its cognitive processes. The second school of thought is rooted in the works on[7] and influenced by social theories of learning (Argyris and Schön 1978; Lave and Wenger 1991; Wenger 1998). This literature defines social learning as a process determined by groups interaction across scales and times (Reed et al. 2010). The aim is a multiparty collaboration embedded in a specific context and leading to specific outcomes. It should rely on a feedback loop between outcomes and context. In social-ecological

[7]Social learning theory in organizational learning literature has been coined under several names such as *situated learning* (Brown and Duguid 1991), as *practice based learning* (Argyris and Schön 1978; Gherardi 2000), and *learning as cultural process* (Cook and Yanow 1993; Yanow 2000).

systems, the context of social learning includes the governance structure (all the actors, institutions and stakeholders) and the environment; practices that facilitate flows of ideas and resources are essential factor for adaptability. The governance structure has a strong influence on the nature of multiparty cooperation because it influences access to information that can impede social learning. The multiparty interactions regard two different aspects: (1) the processing of factual information about a problem and (2) the solution of the problem (Pahl-Wostl et al. 2007a). The issue of management is a core aspect of participatory processes because it makes the problem definition, the direction setting, its implementation, the type of ground rules and the negotiation strategies chosen and the leadership in the process, explicit and shared. Likewise, the development of shared meaning for a joint action is facilitated by relational practices (Pahl-Wostl et al. 2007a). These may take different forms, such as joint field visits or common training sessions. Benefits of social learning refer both to the measures deployed to deal with problems and to the capacity of the stakeholder group to face problems. Social learning can foster flexibility through an efficient use of resources and cultivating diversity of views. It a resilient perspective, it may be useful to consider the theory of *Communities of Practice* (CoP) developed by Wenger (1998) that explores the links between knowledge, learning and communities within organizations. Communities of practice are people as a concern for something they do and how they learn to do it better as they interact regularly in the process.[8] Wegner emphasizes learning as participative process: learning can be the reason the community comes together or the outcome of members' interactions. Thus, it can be also an unintentional outcome of the multiparty interaction. Not all the communities are communities of practice. Two main elements define a community of practice:(1) the identity is defined by a shared domain of interest; and (2) its members are practitioners (Wenger 2000). In CoP, members are engaged in joint activities and discussions and share information but do not necessarily work together on a daily basis. In the same way, past experiences and ways of addressing recurring problems are recognised as crucial resources.

As mentioned above, communities of practice can be created intentionally with the goal of gaining knowledge or unintentionally; in any case, they take time and sustained interaction. Members learn from each other through sharing information and experiences and developing shared meaning. The communities of practice can be understood as forms used by actors' networks to manage and generate knowledge. This is of particular interest to those fields in which there are different sources of knowledge and a continuous process of learning. As Pahl-Wostl, Craps et al. argue

> this leads as well to a different interpretation of the role of information and the ability of an actor network to use new information in social learning processes [..]. Such understanding of knowledge has implications for the role of information and communication (IC) tools,

[8]Communities of practice can exist in any organization both private and public. The members of these communities are not bound by organizational affiliations; they can span institutional structures and hierarchies. They are crucial to those institutions and organizations that recognize knowledge as a key asset. They fulfil a number of functions regarding the creation, accumulation and diffusion of knowledge. They are a node for the exchange and interpretation of information because they allow information to be moved across organizational boundaries.

which range from simple graphical devices to GIS maps as integrated simulation models[..]" (ibid, 2007, p. 7).

Moreover, a CoP added value concerns the integration of both tacit and explicit knowledge, because it preserves the tacit aspects of knowledge that formal systems cannot capture.

Despite the collaborative nature of social learning processes, strong leadership is required. It does not mean imposing a view, however, there is a need to provide guidance to facilitate and to take away ambiguities due to the complexity of problem domains, asymmetries among the actors in power, resources and expertise members. Hence, the CoP needs a collaborative leadership able to reframe perspectives to shared values and goals. In particular, it is important to determine how actors give meaning to information to categorise the issues in the problem. Differences in contextualising derive from diversity in scientific disciplines and backgrounds. This means that social learning processes are strongly influenced by the governance structure in with actors are embedded, thus rigid hierarchical and bureaucratic structures can become barriers to social learning.

To promote the capacity to create and retain knowledge, institutions should build processes where learning communities can evolve and interact. In doing so, it is necessary to build organizational and technological infrastructures that recognize, support and leverage the learning communities but it even more depends largely on the political will and mobilization.

4.11 The Public Participation

Due to the growing complexity of systems and the increasing role of non-structural measures in tackling risks, new forms of horizontal and vertical collaboration among actors have been arising over the past two decades. This has led to changing the distribution of responsibilities among the parties involved, not only including a multiplicity of public actors but also engaging citizens and the private sector (Kuhlicke and Steinführer 2013).

Traditionally, risk management has had a paramilitary approach (Scanlon 1982). The citizen was considered the core element, who had to be protected, safeguarded from or assisted in a set of adverse events. In the same way, all the activities of PPRR chain were pursued *for*, rather than *with*, the community (Laughy 1991). On the contrary, the adaptive management paradigm of social-ecological resilience recognizes the need to forge new relationships to enhance multidirectional information flows. The concept of the *evolutionary prospective* of resilience also highlights how much social systems (also through individual or collective agency) can adapt or search for alternative development trajectories (Davoudi et al. 2012). This paradigm shift opens risk management out to a new role of the community (Scott 2013).

Citizens, advocacy groups, private organizations, non-profit and public organizations from various backgrounds are part of the system, which adds complexity while providing a larger amount of opportunities for public managers to use in the effective management of emergencies (Kapucu 2012, p. s42).

This means that people at risk are no longer simply exposed to risks, but they can be gradually transformed into active individuals; citizens are no longer objects, but they become new subjects in risk management.

Several issues have contributed to this evolution. First of all, the growing risk perception that is crucial in risk management. Over recent years, there has been a general awareness about the rising number of sources of hazards. An example are floods. In the past, floods had a sole focus on rivers and seas but, in the last decades, new events are continually arising due to climate change and flash floods. Furthermore, there is a more sophisticated view of causality that allows to recognize artificial flood sources and the potential for cascading events that can widen the impact beyond those immediately affected (White 2013).

The Information Age is a contributing factor to this awareness; today people can access information, knowledge and news easily. At the same time, this does not mean only benefits. The large amount of information, in fact, contributes also to increase the uncertainty in understanding real risks; furthermore, media can contribute to falsifying reality, emphasizing drama beyond scientific evidence. In addition, it should be stressed that the information about risks that institutions share with the public is related to the credibility and trust that citizens attribute to those institutions (Baggett et al. 2006). Lastly, the different attention in the PPRR chain can also influence risk perception. For instance, Swedish society trusts more in the prevention stage to avoid the emergency; on the contrary, the Italian society is more prepared in emergency response rather than in prevention.

A second aspect has contributed to citizens participation; it is the emerging interest of citizen groups to be engaged in disaster planning. It initially emerged with the risks associated to nuclear power during the 1970s, and it has expanded to include environmental issues such as natural hazards and disasters (Quarantelli 2000). People have started to have a proactive attitude in handling risks, searching information and warnings of danger. Bottom-up activities of communicating advice, education and information are continuously growing.

Finally, but maybe most importantly, there is a growing willingness of people to act collaboratively with authorities. Bottom-up experiences in which people share information on different platforms are on the increase. In particular, there is a shift from an *emotional* participation (persons who upload photos and videos during disastrous events), to a *conscious* participation, where information is shared with a community to solve a problem.

The spread of Web 2.0 in which users become active participants rather than observers has largely contributed (Pearce 2003). Web 2.0 has suggested a reconceptualised World Wide Web (O'Reilly 2005; Hagemann et al. 2007), characterized by a decentralized mode of production, role of servers as accumulators of content from distributed sources, and exchange of content among users (Surowiecki 2004).

Similar evolution has characterized cartography. Usually, reading a map requires some training but today the diffusion of Google Maps and navigators make users more familiar with maps and geographical data than before. The increase in geographical applications has improved people's capability to understand information and has hopefully reduced the number of *map-blind* people with severe problems in understanding what maps describe. The Google map tools let the user know and perceive physical aspects of land in a real image that is easy to understand.[9]

At the same time, nowadays data is or could be acquired easily from citizens or through collaborations between citizens and authorities; there is an impressive spread of geographic information creation, sharing, dissemination, and use in the form of user-generated web content (crowd-sourced spatial data) (Elwood 2010).

Today, several Apps allow to images, graphics and maps to be transmitted directly to users. Many experts think geographical information science has much to offer to the development of collaborative technologies in terms of managing teams' knowledge and coordinating teams' actions (Cai et al. 2005). For example, sharing information, by geographical information system support, can be used to assist people in case of danger or accident in choosing an alternative route to their destination.

When the public is at risk of real or potential threats, treatment options may be limited, direct interventions may take time to be organised and resources may be few. In this context, more widespread communication could provide individuals and communities with information needed to survive an emergency.

Similarly, people may act collaboratively with authorities. Web and mobile platforms are of increasing importance for data collection. They help greatly in collecting accurate geospatial data, as well as delivering helpful location-aware applications that can be easily used by everyone.

New trends show how social networking is supposed to improve interactivity between a local government and the public. In some cases, the local government can release information to the population; in other cases, citizens create and share geographical information with other people or authorities following a collaborative approach. However, the main problems in using data provided by common people are data quality, accuracy and integrity. Also, the process of involving people should be an interactive learning process which needs to take into consideration past experiences, feedback loops and interaction with others.

4.12 Moving Towards a More Desirable Trajectory

Social-ecological resilience can be used as a conceptual framework for re-thinking risk management in a more dynamic and holistic way. While it is true that persistence is a crucial dimension of risk management, some other issues are commonly over-

[9]At the same time, we should be aware of the limits of web cartography diffusion. Usually, people that use freely available geo-web tools are much more focused on where to go and how to get there, while the question of "what is it?" and "which are the components of the land?" are not considered due to the lack of interpretative information provided by the web. A deeper knowledge of the landscape elements requires a thorough interpretation of the objects in an area and a merging with data from already measured roads, catastrophes, vegetation, soil, map (Salvemini et al. 2011).

looked. These concern mostly the human action and the capacity of the governing bodies.

Emergency management, indeed, is affected by a deep complexity in terms of uncertainty, number of stakeholders involved, extension in space and time and potential damages. Managing an emergency means identifying and contextualizing the crisis, taking decisions under pressure, organizing individuals and sources, but also improving management skills from one crisis to the next, as well as studying and working to avoid or mitigate new crises (Koraeus 2008). Therefore, besides persistence it requires adaptation to change, preparedness, flexibility and rapidity to find alternative paths in facing problems. In resilience literature these abilities are strongly interconnected to governance institutions and their learning capacity. But how this challenge can be fostered and addressed?

As it was mentioned, in practice, problems do not present themselves as given, but they require an amount of different knowledge coming from professional pluralism that characterizes the PPRR chain. What stands out is a cross-scale claim characterized by interactions of scales and levels in the knowledge system, where the term *knowledge* refers to the process of interpreting and understanding data that should support actors in taking the right decisions.

Working on knowledge creation, implementation and sharing could be a way for improving the quality of public engagement and consequently the system's resilience. This means not only an enlargement in the use of technology but it especially implies a greater attention to people and processes that requires time and political will. It should build communities of interest that step outside the boundaries of existing governance structure but that can critically assess the scales and time of interventions. In the next chapter, it will be analysed a case study in which the reconfiguration of the knowledge sharing infrastructure was the opportunity to straighten the cooperation among stakeholders (including new actors and citizens), to activate resourcefulness and to introduce innovations into risk management processes.

References

Adger WN (2006) Vulnerability. Glob Environ Change 16:268–281

Ambrosini V, Bowman C (2001) Tacit knowledge: some suggestions for operationalization. J Manag Stud 38:811–829

Anderies JM, Janssen MA, Ostrom E (2004) A framework to analyze the robustness of social-ecological systems from an institutional perspective. Ecol Soc 9:18

Argyris C, Schön DA (1978) Organizational learning; a theory of action perspective. Wesley, Reading

Baggett S, Jeffrey P, Jefferson B (2006) Risk perception in participatory planning for water reuse. Desalination 187:149–158

Bandura A (1977) Self-efficacy: toward a unifying theory of behavioral change. Psychol Rev 84:191

Berkes F, Colding J, Folke C (2003) Navigating social-ecological systems: building resilience for complexity and change. Cambridge University Press, Cambridge

Bignami DF (2010) Protezione civile e riduzione del rischio disastri. Metodi e strumenti di governo della sicurezza territoriale e ambientale. Maggioli Editore

Bowditch JL, Buono AF (2005) A primer on organizational behavior. Wiley, New York
Brown JS, Duguid P (1991) Organizational learning and communities-of-practice: toward a unified view of working, learning, and innovation. Organ Sci 2:40–57
Buizer M, Arts B, Kok K (2011) Governance, scale and the environment: the importance of recognizing knowledge claims in transdisciplinary arenas. Ecol Soc 16:18
Cai G, MacEachren AM, Brewer I et al (2005) Map-mediated geocollaborative crisis management. In: Intelligence and security informatics. Springer, Berlin, pp 429–435
Cash DW, Adger WN, Berkes F et al (2006) Scale and cross-scale dynamics: governance and information in a multilevel world. Ecol Soc 11:8
Chapin FS, Lovecraft AL, Zavaleta ES et al (2006) Policy strategies to address sustainability of Alaskan boreal forests in response to a directionally changing climate. Proc Natl Acad Sci 103:16637–16643
Comfort L, Ko K, Zagorecki A (2004) Coordination in rapidly evolving systems: the role of information. Am Behav Sci 48:295–313
Cong X, Pandya KV (2003) Issues of knowledge management in the public sector. Electron J Knowl Manag 1:25–33
Cook SD, Yanow D (1993) Culture and organizational learning. J Manag Inq 2:373–390
Costanza R, Wainger L, Folke C (1993) Modeling complex ecological economic systems. Bioscience 43:545–555. https://doi.org/10.2307/1311949
Crosta P (1998) Politiche. Quale conoscenza per l'azione territoriale. FrancoAngeli, Milan
Davoudi S, Brooks E, Mehmood A (2013) Evolutionary resilience and strategies for climate adaptation. Plan Pract Res 28:307–322
Davoudi S, Shaw K, Haider LJ et al (2012) Resilience: a bridging concept or a dead end? "reframing" resilience: challenges for planning theory and practice interacting traps: resilience assessment of a pasture management system in northern Afghanistan urban resilience: what does it mean in planning practice? Resilience as a useful concept for climate change adaptation? the politics of resilience for planning: a cautionary note. Plan Theory Pract 13:299–333
Easterby-Smith M, Lyles MA (2003) The Blackwell handbook of organizational learning and knowledge management. Blackwell Pub., Malden
ECHO (2014) EU civil protection legislation - ECHO factsheet
Elwood S (2010) Geographic information science: emerging research on the societal implications of the geospatial web. Prog Hum Geogr 34:349–357
European Communities (2009) Communication from the commission to the european parliament, the council, the european economic and social committee and the committee of the regions—a community approach on the prevention of natural and man-made disasters
Folke C (2006) Resilience: the emergence of a perspective for social–ecological systems analyses. Glob Environ Change 16:253–267
Folke C, Carpenter S, Elmqvist T et al (2002) Resilience and sustainable development: building adaptive capacity in a world of transformations. AMBIO J Hum Environ 31:437–440
Folke C, Carpenter SR, Walker B et al (2010) Resilience thinking: integrating resilience, adaptability and transformability. Ecol Soc 15:20
Gherardi S (2000) Practice-based theorizing on learning and knowing in organizations. Organization 7:211–223
Gibson CC, Ostrom E, Ahn TK (2000) The concept of scale and the human dimensions of global change: a survey. Ecol Econ 32:217–239
Gunderson LH, Holling CS (2002) Panarchy: understanding transformations in systems of humans and nature. Island, Washington
Hagemann S, Letz C, Vossen G (2007) Web service discovery-reality check 2.0. In: Third international conference on next generation web services practices, 2007. NWeSP 2007, pp 113–118
Hendriks PH, Dessers E, Van Hootegem G (2012) Reconsidering the definition of a spatial data infrastructure. Int J Geogr Inf Sci 26:1479–1494
Kallinikos J (2001) The age of flexibility: managing organizations and technology. Academia Adacta, Lund, Sweden

Kapucu N (2012) Disaster and emergency management systems in urban areas. Cities 29:S41–S49
Kok K, Veldkamp TA (2011) Scale and governance: conceptual considerations and practical implications. Ecol Soc 16:23
Koraeus M (2008) Who knows? The use of knowledge management in crisis. CRISMART, National Defence College
Kuhlicke C, Steinführer A (2013) Searching for resilience or building social capacities for flood risks? Plan Theory Pract 14:103–140
Langefors B (1996) Essays on infology. Chartwell-Bratt
Laughy L (1991) A planner's handbook for emergency preparedness. Centre for human settlements. University of British Columbia, Vancouver
Lave J, Wenger E (1991) Situated learning: legitimate peripheral participation. Cambridge University Press, Cambridge
Lindberg H, Sundelius B (2012) Whole of society disaster resilience: the swedish way. In: The McGraw-Hill homeland security handbook. strategic guidance for a coordinated approach to effective security and emenrgency management, 2nd edn. McGraw-Hill, New York, pp 1295–1319
McEvoy D, Fünfgeld H, Bosomworth K (2013) Resilience and climate change adaptation: the importance of framing. Plan Pract Res 28:280–293
Nonaka I (2008) The knowledge-creating company. Harvard Business Review Press
O'Reilly T (2005) What is web 2.0
Pahl-Wostl C (2009) A conceptual framework for analysing adaptive capacity and multi-level learning processes in resource governance regimes. Glob Environ Change 19:354–365
Pahl-Wostl C, Craps M, Dewulf A, et al (2007a) Social learning and water resources management. Ecol Soc 12:5
Pahl-Wostl C, Sendzimir J, Jeffrey P et al (2007b) Managing change toward adaptive water management through social learning. Ecol Soc 12:30
Palermo PC, Ponzini D (2014) Inquiry and design for spatial planning. Three approaches to planning research in late modern cities. In: Research methods in spatial and regional planning. Routledge
Pearce L (2003) Disaster management and community planning, and public participation: how to achieve sustainable hazard mitigation. Nat Hazards 28:211–228. https://doi.org/10.1023/A:1022917721797
Pede E (2018) Improving resilience through cross-scale knowledge sharing. In: Brunetta G, Caldarice O, Tollin N et al (eds) urban resilience for risk and adaptation governance. Theory and practice. Springer, Berlin
Polanyi M (2009) The tacit dimension. University of Chicago Press, Chicago
Quarantelli EL (2000) Disaster planning, emergency management and civil protection: the historical development of organized efforts to plan for and to respond to disasters. Preliminary, p 301
Reed MS, Evely AC, Cundill G et al (2010) What is social learning? Ecol Soc 15:11
Renn O (2015) Stakeholder and public involvement in risk governance. Int J Disaster Risk Sci 6:8–20. https://doi.org/10.1007/s13753-015-0037-6
Resilience Alliance (2010) Assessing and managing resilience in social-ecological systems: a practitioner's workbook, Version 2.0
Rogers K (1998) Managing science/management partnerships: a challenge of adaptive management. Conserv Ecol 2:R1
Rothe D (2017) Climate change and security: from paradigmatic resilience to resilience multiple. In: Chandler D, Coaffee J (eds) The Routledge handbook of international resilience. Routledge, New York
Roux DJ, Rogers KH, Biggs HC, et al (2006) Bridging the science–management divide: moving from unidirectional knowledge transfer to knowledge interfacing and sharing. Ecol Soc 11:4
Salvemini M, Vico F, Iannucci C (2011) Plan4all project. Interoperability for spatial planning. Plan4all Consortium
Sapountzaki K, Wanczura S, Casertano G et al (2011) Disconnected policies and actors and the missing role of spatial planning throughout the risk management cycle. Nat Hazards 59:1445–1474
Scanlon J (1982) The roller coaster story of civil defence planning in Canada. Emerg Plann Dig 6

Schön D (1971) Beyond the stable state. Random House
Schön DA (1982) Some of what a planner knows a case study of knowing-in-practice. J Am Plann Assoc 48:351–364
Schön DA (1983) The reflective practitioner: how professionals think in action. Basic Books, New York
Scott M (2013) Living with flood risk. Plan Theory Pract 14:103–140
Smith K (2009) Environmental hazards: assessing risk and reducing disaster, 5th edn. Routledge, New York
Surowiecki J (2004) The wisdom of crowds: why the many are smarter than the few and how collective wisdom shapes business, economies, societies and nations. Doubleday, New York
Termeer CJAM, Dewulf A, Lieshout M van (2010) Disentangling scale approaches in governance research: comparing monocentric, multilevel, and adaptive governance. Ecol Soc 15:29
Tsoukas H (2005) Complex knowledge: studies in organizational epistemology. Oxford University Press, Oxford
Walker BH, Carpenter SR, Rockstrom J et al (2012) Drivers, "slow" variables, "fast" variables, shocks, and resilience. Ecol Soc 17:30
Wenger E (1998) Communities of practice: learning, meaning, and identity. Cambridge University Press, Cambridge
Wenger E (2000) Communities of practice and social learning systems. Organization 7:225–246
White I (2013) The more we know, the more we know we don't know: reflections on a decade of planning, flood risk management and false precision. Plan Theory Pract 14:103–140
Wilbanks TJ (2006) Chapter 2 how scale matters: some concepts and findings. In: Bridging scales and knowledge systems, pp 22–35
Wilbanks TJ, Kates RW (1999) Global change in local places: how scale matters. Clim Change 43:601–628
Yanow D (2000) Seeing organizational learning: a cultural' view. Organization 7:247–268
Young O (2006) Vertical interplay among scale-dependent environmental and resource regimes. Ecol Soc 11:27–43
Zaucha J, Davoudi S, Slob A et al (2016) State-of-the-lagoon reports as vehicles of cross-disciplinary integration. Integr Environ Assess Manag 12:690–700

Chapter 5
Pathways to Operationalise: Umbria Region Case (Italy)

Abstract This chapter unpacks the experience of the Italian region Umbria in facing seismic risk, according to the notion of resilience. This provides how resilience has been variously imagined, shaped and transformed in terms of discourses, values and actions in a practical experience. Umbria is one of the most seismic areas in the centre of the country. After the 1997 earthquake, the authorities started a long-term process, firstly, of re-thinking the response activities of civil protection, and, subsequently, of how the territory relates to a natural hazard in term of capacities of resilience. The reshape started with an organisational innovation aimed at improving the acquisition and transfer of knowledge. Over the years, it has caused the growth of social interaction of involved actors and has led to the reorganisation of governance structures and institutions accordingly to make them more responsive and relevant in the PPRR chain. The results are not limited to civil protection and the response phase, but also affect other sectors, in particular spatial planning. Finally, the call for resilience has also involved the community that, on one hand, wants to be more informed and, on the other, has become part of the response system.

Keywords Umbria · Italy · Emergency management · Knowledge system · Explicit and tacit knowledge · Learning process

5.1 A Fragile Territory: Italy

Italy is one of the most densely inhabited European territories, and one of the countries at greater risk for natural disasters. The country is prone to natural hazards of many kinds with vast economic, social and environmental costs. Its vulnerability is amplified by high density, weak enforcement of building restrictions, the extensive historical heritage and the increasing abandonment of many territories (Bignami 2010; Meloni 2015).

Hydro-geological risks are the most frequent and the damages caused results both in terms of economic costs and deaths. For instance, over the last 50 years (1966–2015), landslides and floods have caused 1,947 deaths, 69 missing and 2,534 injured (ISPRA 2018).

E. Pede, *Planning for Resilience*, SpringerBriefs in Geography,
https://doi.org/10.1007/978-3-030-17262-6_5

Table 5.1 List of major earthquakes in Italy between 1900–2018

Data	Region	Magnitude	Deaths
1908	Messina (Sicily)	7.2	75,000–200,000
1915	Avezzano (Abruzzo)	7.0	29,978–32,610
1930	Irpinia	6.6	1,404
1976	Friuli	6.5	900–978
1980	Campania, Basilicata	6.9	2,483–4,900
1990	Sicily	5.6	19
1997	Umbria	6.4	11
2002	San Giuliano di Puglia (Molise)	5.9	30
2009	L'Aquila (Abruzzo)	6.3	309
2012	Emilia-Romagna	5.8	20
2016	Lazio, Marche, Umbria	6.2	299

Source Own elaboration on data of INGV (https://emidius.mi.ingv.it)

Likewise, numerous seismic faults cover the territory. Up to 40% of the Italian population is estimated to live in highly seismic areas (zone 1 and 2)[1] where 60% of buildings are not constructed according to seismic codes (OECD 2010). The most exposed areas are the Apennine Mountains, the Calabrian Arc, Eastern Sicily and in the north east of the Friuli Region. Furthermore, most of the territory is affected by multiple risks that can create cascade effects. For instance, three active volcanoes (Etna, Stromboli and Vesuvius) contribute to increase the seismic risk; one of which, Vesuvius, is estimated as the most densely populated volcanic region in the world.

Natural risks, more than elsewhere, determinate high exposure not only due to the geological and hydrogeological features but, even more, to its historical polycentric structure of settlement characterized by a dense urban network. However, the earthquakes that have occurred in the last 30 years have shown the low resilience of the territories involved. Despite the forecasting of the seismic hazard through the recognition of the seismic-prone areas, indeed, the number of deaths and building damages still registers substantial losses (see Table 5.1).

Largely driven by emotional reactions in response to disasters,[2] Italian risk management system first started to evolve in the 20th century and has continued to do so

[1]The first laws in this field were established in 1971 and 1974. Nowadays the Italian territory is zoning in 4 zones according to their seismic risks classified by the Peak Ground Acceleration (PGA) and by the frequency of events. The classification is continuously updating because there are still analyses in progress in all the country. Each zone is divided in subzones for a better evaluation at different scales. Nowadays the Italian territory is composed of 175 seismic districts (sea areas are also included), characterised by high or medium risk. These subdivisions allow to identify the different earthquakes and organise the rescue activities.

[2]Italy is not the only country where the evolution of the emergency system is a consequence of calamitous events (Bignami 2010). For instance, the EU directives related to chemical and industrial risks and to hydrogeological risk were issued in response to accidents.

up into the present For this reason, in the Italian prevision-prevention-response and recovery chain (PPRR chain, see Sect. 4.2), attention has been mostly focused on extraordinary interventions and emergencies with a limited time horizon, while prevention and prevision have been underestimated or limited to land use or to structural resistance of building and infrastructure to threats of hazards (Bedini and Bronzini 2018). This lack of importance affects, in particular, the transition between response and prevention in which there has never been a real reflection on the relationship between planning and seismic risk, but only a bounce back to normalcy (Galuzzi 2014; Alexander 2018a).

Examining the literature on risk management, even the research appears to be polarized between the engineering- technical side and the social studies side (Menoni 2014).[3] The former is dominated by hazard analysis, building technical structural component and, especially in the last decade, early warning systems to limit cascading effects. The latter, instead, has been mostly focused on risk perception, the impacts of disaster on the communities and the preparedness of citizens (Mela et al. 2017).

Despite the increasing interest in disaster studies, few projects and researches have considered the comprehensive approach that risk management requires—which is to recognise the importance of cultural, social and economic aspects in response and recovery (Di Giovanni 2016). The low priority given to the implication of planning in risk management resolutely affects the general effectiveness of PPRR chain; in Italy, this aspect is even more critical. For instance, the absence of managerial transparency and the widespread corruption in procurement tends to cause a waste of time in reconstruction; this can last decades without a reflection on the conditions that had co-determined the breakdown of the system in the first place.[4]

The present chapter explores how an evolutionary resilience approach (see Chap. 4) can be introduced in a system that has traditionally had an engineering interpretation of risk management (Campos Venuti 2014; Oliva 2014). It critically examines the approach used to deal with the aftermath of the 1997 earthquake in the Umbria region. Although not as well known, this case is considered to be a milestone because it represents a new approach to risk management in Italy, which has now led the way to a comprehensive rethinking of the process of interaction among multiple levels of organizations. The process has amplified the approach to risk management from the dimension of persistence to the concepts of adaptability and transformability across multiple scales and time frames (Davoudi et al. 2016). It has succeeded in this doing, especially through a closer connection between risk management and spatial planning based on knowledge sharing, social learning and stakeholder engagement.

This chapter will, firstly, give a concise reconstruction of the normative framework about risk management, spatial planning and the allocation of responsibilities in Italy. Successively, it will explore the process that has been set in Umbria region:

---

[3]Over the last years, a new topic is increasing its popularity, disaster insurance.

[4]It must be said that the long term of reconstruction also impacts on academic production. Therefore, scholars give a limited consideration to reconstruction and recovery phases by favoring first emergency phase. The result is a lack of dissemination of the planning practices in recovery and reorganization of the urban structure and risk management chain that some territories are experiencing (Menoni 2017).

the organizational innovation, the connection between critical sectors, the balance between planned and self-organized actions and the stakeholder engagements. The aim is to illustrate the transformation of a system with a formally rigid structure to a more comprehensive and integrated one. Building resilience not only involves planning for shock recoveries, but also cultivating stakeholders' involvement, which takes effect by integrating the diverse disciplines and linking these into the policy process.

5.2 The Institutional Framework in Italy

The legal framework and the political-administrative system significantly determine how flexible the risk management structure is and how inclined it is to be not only comprehensive, but also integrated and subsequently what the obstacles are to strengthen the resilience of a community.

In the last two decades, five earthquakes of considerable size have struck Italy (Umbria and Marche 1997, San Giuliano di Puglia 2002, L'Aquila 2009, Emilia-Romagna 2012, Lazio-Marche-Umbria 2016) causing 669 deaths.[5] Over time, both legislation and practices have sought to improve the approach to disasters with mixed results: successful in the cases of Umbria, Marche and Emilia-Romagna, decidedly less so in the historical centre of L'Aquila.

In order to analyse the organizational structures of the PPRR chain, attention has to be paid to how long-term and short-term measures are combined, in the integration of emergency management measures with spatial planning and territorial strategies (Prenger-Berninghoff et al. 2014).

In general, the Italian risk management approach tends to be disaster *reactive*, meaning that it is based on resistance to change (Di Giovanni 2016). Hydrogeological and seismic risk of the territory are, therefore, treated in terms of reduction of exposure and vulnerability, increasing the structural resistance of the buildings and infrastructures, or, when possible, intervening in the unstable areas. There is an overreliance on engineering technologies, standards of conformity to respect and strict regulations, all of which can be partially related to the Italian *urbanism tradition* dominated by a *conformative* model of the planning system (Janin Rivolin 2008). Likewise, some cultural references influence the system. In Italy, the relation between emergency management and spatial planning appears to be strongly focused on emergency management. This attitude is also asserted by the governmental budgets that are largely dedicated towards emergency response and recovery activities rather than spatial planning and risk prevention (Paleari 2018), demonstrating an implicit view that is the more rigid engineering interpretation of resilience.

[5]The earthquake that struck L'Aquila and 57 other municipalities in the Region of Abruzzo on April 6, 2009 had probably the largest media resonance at the international level because its epicenter was in a city of 73,000 people and destroyed most of the historical center and damaged seriously the cultural and monumental heritage (Oliva 2014).

In further detail, the main actor, in regards to emergency preparedness and response, is the *Protezione Civile* (civil protection) while prevention and recovery are mostly driven by spatial planning authorities. The legislation determines which institutions are responsible for each activity and what the requirements are for coordination; this is why a synthesis of the legal framework of civil protection and spatial planning is put forward to better understand the Italian approach to risk cycle.

As in other countries, civil protection has developed from the fire brigade at the local level; but, unlike most European Countries, Italian civil protection involves the whole State organisation, including a civil society that is involved through voluntary organisations. The reasons are rooted in the decentralised territorial system, as well as in the wide range of possible risks.

The National Civil Protection Service (NCPS) consists of the Regions, the Provinces, the Municipalities, the national and local institutions and all other relevant public and private organisations which are designated by law. Law 225/1992 defines civil protection as a structure based on the principle of subsidiarity[6] in which the emergency response depends on the magnitude of disaster and the degree of cascade effects, as well the relationship to the capacity flow of resources, equipment information and knowledge among the involved actors.

Usually, the first level which is responsible for civil protection is the local one, it is the mayor who that has to organise municipal resources according to pre-established plans. This level also has to coordinate civil protection with structural planning, especially, in regard to land use and housing regulations. In case of serious situations, Provinces, Regions and every State administrations, all co-ordinated by the Prefects, have to guarantee their support. This coordination should not be limited to the horizontal interaction among the forces that are involved, but it also need to include the authorities that had previously acted in the forecasting and prevision and prevention steps (Table 5.2).

The main task of the Provinces is to put into action the Provincial Emergency Plan—an organizational plan of the available manpower and equipment—into action. Regions are generally responsible for civil protection, and particularly for risk assessment, emergency forecasting and prevention programmes. In addition, they also have indirect powers in risk management, due to their authority in transportation, education and environment. The national level lays down general guidelines and provides coordination; however, its operative role through the military force is only employed in serious situations.

Additional authorities—fire brigades, police forces, health facilities, public utilities, telephone companies and local volunteer forces—are also involved before and during emergencies, adding complexity to the network of actors.[7] Furthermore, min-

---

[6]Law 225/1992 established the National Civil Protection Service, defined the typology of intervention, responsibilities and the types of activities: prevention, prevision, response and reconstruction.

[7]Depending on the type of disaster, even more institutions and organisations can be involved including: National Alpine Rescue and Speleology Corps, Ministry of Work, Health and Social Policies, ENAC, ENAV (Italian Company for Air Navigation Services, Electrical Power Service Companies), GSE (Gestore Servizi Elettrici), Electrical Power Service Companies Terna, ANAS (Azienda Nazionale Autonoma delle Strade Statali-Auto-routes of Italy) and AISCAT (Associazione Italiana

Table 5.2 Involved actors in the different steps of risk cycle

Actors/Agencies	Prevision	Prevention	Response	Recovery
National Government	x		x	x
Civil protection department	x		x	x
Prefect			x	
Scientific research groups	x			
Region	x	x	x	x
Province	x	x		
Municipality	x	x	x	x
Volunteer service	x	x	x	
Fire-fighters Corp		x	x	

Source Table of the author

istries, universities, public administrations and public research institutes and centres, all have different forms of cooperation, agreements, understandings and working teams with the DCP in order to define both prevention and forecasting measures for the different types of risk.

Finally, volunteer organisations represent an important resource in the Italian emergency management. Indeed, as many as 50% of the personnel involved in an operational response to a disaster may be volunteers.

As already mentioned, the national law 225/1992 and the recent reform of Civil Protection System (L.100/2012) stresses the need of the relationship and coordination between spatial planning and the activities of prevention and emergency management. Despite the fact that spatial planning should be one of the major contributors to the PPRR chain, in Italy its importance is generally rather low. It is mostly put into effect in terms of prohibiting new construction in hazard-prone areas.

As in the case of the civil protection system, the planning legislative framework has been also strongly influenced by the emotional reaction after disasters especially related to landslides and earthquakes. The current national law on urban planning was promulgated in 1942 (L.1150/42) and was modified in 1967 (L.765/1976) as a result of the catastrophic events of 1966.[8] Subsequently, between the '70 s and '90 s, several technical standards for building in seismic zones were introduced along with hazard plans (e.g. basin management plans). This resulted in the management of seismic risk policies that were merely focused on the resistance of buildings or their protection without any correlation among the different parts of an urban area or the

Società Concessionarie di Autostrade e Trafori), State Railway—RFI (Italian Rail Network) and Trenitalia, Fixed line Telephone and mobile phone Companies, RAI (Radiotelevisione Italia), the Italian Postal system, ENI (Ente Nazionale Idrocarburi—Oil and Gas production).

[8]The landslide in Agrigento (Sicily) and the floods in the northern and central Italy.

mutual influences that each element has on the urban system (Olivieri 2004).[9] Only after the earthquake in Umbria and Marche in 1997, and in Molise in 2002, did the anti-seismic prevention extend from single buildings to urban centres, along with the improvement of the national seismic zoning of the territory. This process has led to the guidelines for the seismic micro-zoning that has become to be used, not only as a post-event tool, but also as a prevision tool (Bramerini 2013; Dolce 2013). The Umbria and Marche earthquake signified a crucial point in the relationship between emergency management and planning. It also led to the introduction of *Minimal Urban Structure* (SUM) into urban planning in order to ensure and maintain the economic, social and relational activities after the earthquake (see Sect. 5.7).

Another step forward was achieved through the analysis of *Limit Conditions for Emergency* (CLE) that became compulsory in 2013, after the earthquake in Abruzzo. This sets forth the minimum conditions of accessibility and connection of the territory for the functions that are strategic for emergency operations. Both SUM and CLE[10] encouraged the relationship between civil protection and spatial planning in adopting the systemic resilience approach (Fabietti 2013; Olivieri 2013).

5.3 From Resistence to Comprehensive Approach

Umbria is one of the smallest Italian regions for dimension and population.[11] It is located in the centre of the country and most of its territory is classified within zone 1 and 2 (high and medium hazard level) of the seismic classification hazard. In general, the east part of the region is characterised by high/medium-high seismic phenomena while the west part has a medium-low/low hazard.

On 26 September 1997, at 2.33 a.m., a first foreshock rated 5.7 on the Richter Scale followed by a second one rated at 6.1 at 11.42 a.m. hit the Umbria region. The two foreshocks left enormous damage, especially to the cultural heritage, and 11 dead. The seismic season of 1997, lasting more than six months, forced the region to rethink its system for handling earthquakes, especially by reason of its historical heritage and urban centres. As in most of the Italian earthquakes, the damage suffered by buildings were the main source of losses with consequences on the whole system and its functions.

Since 1998, and in particular after 2006, the Umbria region has started a process for increasing the resilience of the system to seismic events. The challenge has been met through a reorganization of the PPRR chain able to improve the integration of the crucial sectors and to enhance the relationship-building among stakeholders.

---

[9]The concept of prevention based on building resistance was declared in the national law n.64/1974 *Provvedimenti per le costruzioni con particolari prescrizioni per le zone sismiche*.

[10]The Minimal Urban Structure and the Limit Condition for the Emergency are complementary. The first specifically pertains to the planning aspects and the second is mainly linked to the essential emergency conditions.

[11]From the administrative point of view, the Region is composed by two Provinces (Perugia and Terni) and 92 municipalities.

The initial element of this change was the sharing of the different knowledge relating to the multiple sectors (see Sect. 4.7) through a database implementable and accessible by the different authorities. Over the years, the database has also become a bridging tool from emergency to spatial planning. Indeed, the construction of a common conceptual space for the different domains has contributed to the governance with inclusive decision-making processes in the realm of planning, accountability and collaboration. In addition, procedures and studies have been included to improve the urban design and the qualities of urban form against catastrophic events. Umbria region has worked on the idea that overcoming a catastrophic event does not imply only to protect buildings and guarantee their reconstruction, but also to include the urban structure in a more complex, broader operative system, with a multi-scale perspective (SUM, *minimal urban structure*). This approach introduced the concept of complex system and it shows that spatial planning and emergency management are closely related not only for forecasting but also in preventive activities that can have consequences during the emergency (Olivieri 2004, see Sect. 5.7),

Between the summer and autumn of 2016, a new seismic season hit the centre of Italy. On 24 August, the first 6.1-rated foreshock had its epicentre in Accumoli, in the Lazio region, in an area near the borders of the Umbria, Abruzzo and Marche regions. The earthquake injured 365 and caused 297 deaths, mostly located in the epicentre. On 30 October, a new shock of a magnitude rated at 6.6 struck Norcia in the Umbria region. This second shock was the largest earthquake since the 1980s in Irpinia; it mostly caused heavy damage to some structures and to cultural heritage, while no deaths caused by the structural collapse of buildings were recorded. It is hard to draw any comparison between earthquakes because several elements influence the impact of a seismic event; however, the response capacity of Umbria was considered positive compared to the past, especially, thanks to the reshaping process of the disaster risk management cycle and the reconstruction quality after 1997.

5.4 A Bridging Tool: Towards a Territorial Knowledge System

As mentioned above, the project can be summarized as a combination of a technological innovation (*hardware*) and social capital and learning processes (*software*). The hardware is the database called AZIMUT: a geographic information system for developing local and provincial emergency plans able to integrate the essential data and information created by the different authorities involved in the rescue service. It was born of the need to better integrate information and knowledge in order to support decision making and coordinated the actions of the risk cycle. In 1998, the degree of data integration was still poor and most of the plans were available only on paperback; however, the information technology was starting to show its potential. AZIMUT has been designed to be a dynamic tool that is easy to update, be implemented, and be used by different authorities at the same time. It was created with a twofold purpose: to build a unique integrated database and to have a flexible tool in responding to different needs caused by calamitous events or unforeseen situations.

The AZIMUT technology is based on geographical information system (GIS) that enhanced the availability and usability of existing information. During the *peace time*, data are collected to include all the useful elements for forecasting, prevention and emergency response. At the beginning, the project seemed more like an organisational knowledge storage (see Sect. 4.8) to combine the useful data to respond to the sense of urgency of a crisis in which decisions must be made under time pressure.

Over time, in 2012, AZIMUT evolved into a new system called SRD (data collection system).[12] that allows the creation and direction of data georeferencing through the new Data Entry (SRD), which can be consulted by a WebGIS without the need of GIS desktop software. Data sharing via Web-GIS technology provides end-users to simultaneously access multiple data sources that are continuously updated. This can also be used during fieldworks: rescuers can share real-time information (through PDA[13] with wi-fi) with the Municipal Operations Centre about injuries, damages, broken roads, as well as require the intervention of fire fighters and 118 or it can analyse alternative routes to the assembling or meeting points.

Thus, the system has evolved into a dynamic and effective tool for all the phases of the risk cycle. Sharing several data elements in the same space has allowed, indeed, a better planning of the activities from ordinary planning to emergency activities, as well as it has contributed to enhance the relationship between stakeholders improving decision-making. The AZIMUT information system includes a large network of actors: all the Municipalities,[14] the Region, the Prefecture, the Provinces and the Fire Fighters.

The database has been designed not to be a container of information, but to allow municipalities to contribute directly to its implementation in order to have continuous data updates, to increase the quality of information but also to reorient the trajectories of development introducing new knowledge for supporting innovation.

The data flow contributes to identify responsibilities. The system is composed by several parts and actors are providers or applicants according to their own responsibilities. Most of the data are created at local level, but there are also data showing non-local values that are provided by the Provinces (e.g. hospitals, stock areas, etc.) that play a central role in assisting and coordinating the other institutions or also in including other authorities in the network. Finally, the Region, the Prefecture, and Fire Fighters are merely applicants of the information during the events. This allows a flexible use of resources also for different purposes: data can be used at municipal scale, upper scale or can have a cross-scale use at once.

[12]Usually the new system is also called Azimut-SRD.

[13]Personal Digital Assistant, palmtop computer.

[14]The Municipalities are divided into *area manager municipalities* and *afferent municipalities*. The area manager municipalities have the responsibilities of the data integration for all their afferent municipalities. In the Umbria Region, there are only 11 municipalities with more than 20,000 inhabitants, whereas, most of the municipalities have small dimensions and they do not have enough technicians with the right skills to deal with this kind of technology. In these cases, the cooperation with a larger municipality can help the agents of small municipalities to satisfy all the requests.

Despite the initial aim concerned the growing need of up-to-date information during rescue operations, the project was able to evolve into a wider engagement process able to enhance the relationship among stakeholders and to integrate their knowledge to meet the challenges. The added value of this technological innovation was the creation of common meaning of the knowledge among the different stakeholders (policy makers, practitioners and citizens) who interact across the scales. Working towards a common vocabulary was essential to overcome the boundaries between the various discipline involved. It has required a long social learning process. This can be considered as the software of the project, an essential part for the effective operation of the AZIMUT system.

5.5 To Cross Knowledge Boundaries

The main challenge in building the AZIMUT system was the integration of scientific knowledge in order to create an interdisciplinary tool in which various disciplines are involved in understanding and describe the territorial system behaviour against shocks. Besides the organizational innovation, therefore, the experience of AZIMUT was crucial because it allowed to work on network skills of the stakeholders involved. Due to the cultural differences between disciplines, one of the core challenges was to overcome the boundaries between the various disciplines involved (Zaucha et al. 2016). Therefore, a first step in creating the AZIMUT database was to choose which data to integrate defining a common meaning. The priority was given to data that must be included in the Local Emergency Plan (Piano Comunale di Emergenza), the main tool of emergency planning. Its main objectives are: to analyse risk scenarios; to elaborate risk cartographies concerning elements at-risk and useful resources for overcoming the crisis; to provide behavioural guidelines for rescuers and the population. In addition, the plan includes an operative part that identifies the assembling points, the strategic buildings (e.g. hospital, fire stations, etc.) and the resources available during the emergency.

There are different reasons why the Local Emergency Plan was chosen to be the basis of the database. Above all, the plan is compulsory for all municipalities and it must be consistent to regional guidelines; therefore, the structure of the plan—and consequently the required data—are the same for all the municipalities.

At the same time, the plan is elaborated at local scale, but it has several end-users with diverse skills at different levels; consequently, the plan represents a coordination tool in which the problem of *knowledge as a scale* (Cash et al. 2006, see Sect. 4.5) is evident. It simultaneously covers all types of risks, it includes different resources and the knowledge required is characterised by cross-scale interaction, especially in terms of jurisdictional scale as well as in terms of time scale.[15]

[15]The time scale corresponds to the PPRR chain, otherwise, it can also be considered as divided into *before*, *during* and *after* the event.

Having defined the data to insert, a thorough standardisation in collecting data was defined in order to support the collaboration between actors, to improve the manageability of the information and to facilitate data acquisition. Each raw data has had a univocal definition and specific symbols has been established in arrangements between the different scientific disciplines. This has provided the foundation for dialogue building commitment. It has helped to avoid misunderstanding and to strengthen relationship between decision- makers and technicians at different scales.

Even if the network enhancement was one of the aims of the project, only practice has caused a real improvement of the coperative ability of the actors. Initially, the AZIMUT system was mostly a top-down operation drove by a strong political willing and also supported by significant financial resources.

Therefore, the framework laid down by the regional authorities underestimated a few issues, requiring a change in the pathway. For instance, lots of municipalities did not have specific technicians for risk management as well as technicians did not have appropriate informatic expertise. For this reason, experts[16] were engaged in the project to support and improve the skills of the technicians of the municipalities. This long process—that took five years to include all the Municipalities—has contributed to the success of the AZIMUT project. The unexpected barrier in skills showed the institutional willingness and the preparedness to transform and reorient the project looking for solutions to achieve the objective.

At the same time, the social learning process that led to this result has improved the empowerment of the municipalities and has increased confidence among the actors and mutual confidence. Firstly, it has improved the skills of the persons in charge. Secondly, the collection phase and continuous updating has demanded incentivisation to create specific local and provincial offices, dedicated to this task. Finally, the usage of a unique tool throughout the Region—in which data are codified in the same way and with a univocal meaning—has improved the cooperation among people that usually work in different places, but have the same job assignments. This represents a benefit during the emergencies because every technician is trained to work everywhere: in each plan the contents are different, but the working methodology is the same; it means that if a municipality has an emergency it is easier to give support.

Therefore, over the years, the actors' interaction has produced a genuine community of practice (Wenger 2000, see Sect. 4.10). Actually, the community of practice was intentionally planned to gain the knowledge for the Emergency Local Plan and with the support of technological infrastructure; however, its success is based on the trust building process between involved actors that has established vertical and horizontal linkages. It has taken a long time, but it has been crucial for the success of the whole project.

---

[16]The specialists of ANCI Umbria (Association of Italian Municipalities) had the role to support local technicians in data collection and in the use of AZIMUT (ANCI Project for municipalities' support). Each municipality has a referenced specialist of ANCI who helps the municipal technicians to satisfy all the requests of the Province and Region for civil protection.

5.6 Combining Expert and Tacit Knowledge

Further attempts in overcrossing boundaries between the different levels and the various disciplines involved stand out. Besides the codification of explicit scientific knowledge, the project has also contributed to combine the scientific with the tacit knowledge (see Sect. 4.6). The first element that recalls the tacit knowledge concept is the strong link with practice (Schon 1982; Nonaka 2008; Polanyi 2009) that the AZIMUT project mostly aims to improve. In this framework, municipalities have been considered the main holders of knowledge not only because of the better data resolution at local scale, but especially because they are the holders of knowledge due to experience (*tacit knowledge*). Tacit knowledge concerns the collective memory, the knowledge acquired over years by who lives in those territories or by who has had practical experience and expertise in that specific territory during the past emergencies. Especially in small municipalities, people in charge in operational roles have a deep knowledge of the territory that can be integrated to the standard operating procedures.

Likewise, a second element referable to tacit knowledge concerns interpersonal activities. The technical support team have played a key role in the translation and mediation between municipalities and upper levels. They have acted as *mediator* of the scientific and tacit knowledge that is usually held, stored, and perceived differently at different levels. Cash et al. (2006) defined this type of actors as *boundary organisations* or *bridging organisations* because

> they play an intermediary role between different arenas, levels, or scales and facilitate the co-production of knowledge (ibid, pp. 8).

They do not only combine knowledge, but also interpret and make knowledge accessible in the local context. Throughout their role they can synthesise and remove shifts of knowledge between the different levels and scales. They establish functional links within and between levels and facilitate the flow of information and knowledge in the local system. Their role is characterised by: accountability to all stakeholders involved; trust building; translation; coordination and complementary expertise; mediation; leadership.

It must be said that this was an unexpected effect of the project; a lesson learned that allowed to move from single-loop to double and triple-loop learning (Argyris and Schön 1978). Municipalities are more aware now of emergency planning than they were before. The project has improved the technicians' skills, because together, they work, interact and create a new network.

5.7 A New Role for Spatial Planning

As set out in Sect. 5.2, Italy's approach to planning has played a crucial role in reducing vulnerability to natural hazards mainly through the implementation of land-use management and building design codes to regulate development in hazard prone areas. However, limiting the actions at the building scale is not sufficient enough. Urban environments are vulnerable as a system, thus, the structural component of the single buildings has to be considered in relation to the others (D'Amico and Currà 2018). Ideally, the building would be integrated in its block, as well as in the urban structure; likewise, the city should be considered as part of the broader scale.

In this regard, the approach followed by the Umbria region has given important inputs to cross-sector collaboration between emergency and spatial planning that have now been followed in other regions and also influenced the national legislation.

The Umbria regional authority was, indeed, the first region who introduced by law (L.R. 11/2005) the obligation of the *minimal urban structure* in the urban structural plans. The concept of minimal urban structure includes all the functions, itineraries, strategic places and buildings that must be preserved during a crisis in order to best implement the response and the recovery after the event. It considers the interrelations of the elements and evaluates the direct or indirect relationship of vulnerability among the elements and the effects that the fragility of a node could produce in the overall functioning of the whole system.

It is incorrect to think that the minimal urban structure only reduces the vulnerability to risk. Although, at the beginning, the aim of the SUM was mostly related to the management of rescues; over the years, the focus has changed to the other critical elements for the recovery. SUM can, indeed, acts as a strategic tool in preserving the *vitality* of the urban system. Besides the key elements for dealing with the emergencies (main communication routes and nodes, the escape routes, the secure open and closed spaces, the key buildings), SUM includes those aspects that are crucial for resuming the economic, social and relational activities (public services, relevant economic activities and also the places where the local community identifies itself with), after the earthquake (Corradi and Fabietti 2017).

Generally, one of the indirect consequences of earthquake is the depopulation of the territories and the economic losses in term of employers and number of enterprises; this is mostly due to the long amount of time it takes for reconstruction (Alexander 2018b). For instance, according to ISTAT (www.istat.it), the Italian national statistical agency, the registered, resident population of L'Aquila city was 72,696 on 31 December 2009 and only 69,439 on 31 December 2018.

In some Italian regions, the introduction of SUM has represented the basis for promoting recovery strategies able to reprogram the development trajectories of the territories struck by catastrophic events. For instance, the use of SUM has influenced the successful response to the aftermath of the 2012 earthquake in Emilia Romagna[17] and the dynamic economy that characterises the was preserved (Bedini and Bronzini 2018).

Likewise, SUM can be highly suitable for sparsely populated areas characterised by polycentric system. In these cases, SUM is used at urban and territorial levels

---

[17]The experience in Emilia-Romagna represents the most Italian successful case in dealing with the earthquake. It was developed from a complex bottom-up process for relaunching the a very large territory (Bedini and Bronzini 2018).

involving several municipalities with the scope to ensure services in collaboration and avoiding the abandonment of the territories.

Finally, SUM cannot be fixed in time, but has to remain flexible, able to adapt according to the transformation of the trajectories of development. In this sense, it should not only be viewed as part of the structural planning aimed to improve the robustness of the system, but also utilised as a monitoring and assessment tool for the development strategies of the planning (Corradi and Fabietti 2017; Menoni 2017). It is for this reason why it is important to understand by which means SUM intends to connect and to influence its policies (e.g. structural plan, local emergency plans but also strategic plans). Planning should not limit its role to prevision and prevention, rather, in order to be most effective, it must also expand its flexibility to facing hazards.

5.8 The Community Engagement

Working on urban vulnerability also means acting on all the elements of the urban system including citizens; people have their own dynamism that can influence the resilience of the system (Kapucu 2012; Kuhlicke and Steinführer 2013; Pearce 2003). In Italian law 112/1998, the involvement of the population is required but usually the concept of *commander and control*—in which people are passive actors to take care of—is still in use. Furthermore, the most common actions about citizens' participation in rescue services concern education and information activities, especially in schools or for public officials and technicians.

In this field, the Umbria region has introduced important innovations toward the proactive role of citizens. Beside the reorganisation of the PPRR chain, two projects have contributed to implementing citizens' participation.

Concerning information activities, beside the education projects in schools, activities on the web have been introduced. Social networks and mobile apps have several benefits: they are channels that customers can access at a time and place of their choice, which they easily understand and are comfortable with, which can offer a more open and transparent relationship between customers and authorities offering new ways of access to information. The information can be useful for citizens in ordinary time because they can have a better knowledge of the risks of the territory where they live, as well as during emergencies.

Through a website and an app for mobiles, citizens have access to the information related to a specific area and can receive alerts of risks present in the zone of interest. The online platform is used to share the emergency local plan, where data is made available for locals showing the main information of the Emergency Local Plan structured in layers showing points (e.g., assembling points) linear (e.g. roads with flood risk) and polygonal zones (e.g. hydraulic hazard areas). It also shows the territorial information about each municipality and the news about risks, reporting the daily Regional Civil Protection centre warning on risk levels for each zone. Finally, it includes videos and links on the weather forecast and educational material.

Likewise, the Region has also experienced an innovative methodology for boosting the proactive role of citizens and the image of the *citizen as the first rescuer*,[18] in order to increase the awareness of citizens about the degree of vulnerability of the urban system and/or allows them to act as first rescuers.

The concept of *citizen as first rescuer* aims to improve the self-protection measures of the community. The main aim is to increase the sense of responsibility of the population and proactive behaviours, which vary in range from awaiting rescue to a leading role in action. This innovation is based on the collaboration of selected citizens (called *capozona)* that have to ensure a safe and proper evacuation of the population in their assigned area. The capozona also has the responsibility to share information (through PDA with wi-fi and connection to the Azimut geodatabase) with the Municipal Operations Centre. In this way, the Municipal Operations Centre can receive and give real-time information in terms of injuries, damages, broken roads, as capozona can require intervention of fire fighters and the 118 service.

This project has shown the potential of the web and smartphones in the rescue field. However, it is important to underline that such innovations have many problems in terms of data quality and accuracy. Nevertheless, this methodology has shown the potential application of these technologies in small municipalities with few inhabitants, whereas, it is more difficult to support the same approach in a city. This is a limit but at the same time, an opportunity for marginalised areas, in which rescue activities are slower and more difficult and where the community is used to having better interaction.

5.9 Elements for a Resilient Approach

This chapter attempts to apply the social-ecological and evolutionary resilience concepts to the experience of the Umbria region, in Italy. As discussed in Chaps. 3 and 4, within this framework, resilience is not only about persistence, robustness and the return to normalcy; but also about adapting to change and, more importantly, being prepared to change trajectories of development or ways to act (Davoudi et al. 2016).

Seismic risk management is characterized by many interdependencies, the diversity of stakeholder interests, and multiple views of the problem. To deal with this complexity, we need a better integration of the diverse scientific knowledge, the stakeholder involvement and social and political mobilization (see Chap. 4).

As illustrated in Sect. 5.2, the Italian legislation system is still suffering from a fuzzy framework, characterized by an overreliance on standards of conformity and regulations that make the risk management system disaster *reactive*, i.e. one that is based on resistance to change.

---

[18]The trial of this method, called S.I.S.M.A. project (System Integrated for Security Management Activities), was part of the INTERREG-CADES in 2010.

However, after the 1997 earthquake the Umbria region initiated a process of rethinking, which introduced technological and process innovations that can be associated with the principles of persistence, flexibility and preparedness of evolutionary resilience (see Sect. 4.1).

Persistence remains essential in all the resilience interpretations, even when we refuse a simple engineering interpretation of the concept (see Chap. 3). Persistence must, however, be accompanied by flexibility of choice over the crucial features of the system to be made robust according to the vision of development that a community has chosen for itself (Davoudi et al. 2016). In the Italian system, the introduction of SUM can be regarded as a step towards this approach. With SUM the focus on robustness has been changed from the *building scale* to the *systemic scale* in which relations between elements are as important as the building resistance. Furthermore, priority is given, not only to buildings and infrastructure, but also to functions and identity. This change of paradigm has led to deeper integration between emergency planning and ordinary planning. The key challenge of governing this integration is how to promote networking and, especially, how to overcome the barriers in communication between different disciplines.

Initially, the Umbria region made use of organisational concepts and arrangements that promoted knowledge creation and knowledge transfer that were typical of the knowledge management approaches (Easterby-Smith and Lyles 2003, see Sect. 4.8), such as the standard definition of the data to be collected and the strict definition of data providers and users. These days, the database AZIMUT provides the backbone of the project, but its success has depended on the quality of the public engagement and the process of social learning in which stakeholders have been involved. This approach has demonstrated that working towards a common meaning is essential to develop a shared understanding. Barriers to knowledge sharing and communication have been identified in the process and discussions made to eliminate them. This has been the most difficult challenge of all, as they have been resources intensive and time consuming and require a high degree of social learning, drawing on all forms of knowledge including the tacit knowledge embedded in collective memory. In the meantime, preparedness has developed, not only in term of awareness of risks and their related preventative and responsive actions, but also in terms of building trust among stakeholders and through a participatory process that has also involved the community.

The result has been a knowledge management system that combines human resources (intellectual capital, explicit and tacit knowledge and social learning), technological aspects (digital data storage and retrieval system, policies and procedures) and urban forms that greatly enhance the efficiency and effectiveness of resilience in the region.

References

Alexander D (2018a) Civil protection in Italy-coping with multiple disasters. Contemp Ital Polit 10:393–406

Alexander DE (2018b) L'Aquila, central Italy, and the "disaster cycle", 2009–2017. Disaster Prev Manag Int J

Argyris C, Schön DA (1978) Organizational learning; a theory of action perspective. Wesley, Reading

Bedini MA, Bronzini F (2018) The post-earthquake experience in Italy. Difficulties and the possibility of planning the resurgence of the territories affected by earthquakes. Land Use Policy 78:303–315

Bignami DF (2010) Protezione civile e riduzione del rischio disastri. Metodi e strumenti di governo della sicurezza territoriale e ambientale. Maggioli Editore

Bramerini F (2013) Una strategia di prevenzione del rischio sismico per gli insediamenti. Urban Doss 130:5–7

Campos Venuti G (2014) Terremoti, Urbanistica e Territorio. 154:53–68

Cash DW, Adger WN, Berkes F et al (2006) Scale and cross-scale dynamics: governance and information in a multilevel world. Ecol Soc 11:8

Corradi E, Fabietti V (2017) La ricostruzione come metodo, Cosa insegna la storia recente degli eventi sismici in Italia. Urban Inf 272:773–777

D'Amico A, Currà E (2018) Urban resilience in the historical centres of Italian cities and towns. Strategies of preventative planning. TECHNE-J Technol Archit Environ 257–268

Davoudi S, Zaucha J, Brooks E (2016) Evolutionary resilience and complex lagoon systems. Integr Environ Assess Manag 12:711–718

Di Giovanni G (2016) Cities at risk: status of Italian planning system in reducing seismic and hydrogeological risks. Tema J Land Use Mobil Environ 9:43–62

Dolce M (2013) Politiche di prevenzione del rischio sismico a scala nazionale. Urban Doss 130:3–4

Easterby-Smith M, Lyles MA (2003) The Blackwell handbook of organizational learning and knowledge management. Blackwell Pub., Malden

Fabietti V (2013) Dalla CLE alla SUM: i contenuti urbanistici della protezione dai rischi. Urban Doss 130:38–39

Galuzzi P (2014) I rischi e la cura. Urbanistica 154:4–7

ISPRA (2018) Dissesto Idrogeologico in Italia: pericolosità e indicatori di rischio. Roma

Janin Rivolin U (2008) Conforming and performing planning systems in Europe: an unbearable cohabitation. Plan Pract Res 23:167–186

Kapucu N (2012) Disaster and emergency management systems in urban areas. Cities 29:S41–S49

Kuhlicke C, Steinführer A (2013) Searching for resilience or building social capacities for flood risks? Plan Theory Pract 14:103–140

Mela A, Mugnano S, Olori D (2017) Territori vulnerabili: verso una nuova sociologia dei disastri italiana. Franco Angeli, Milano

Meloni B (ed) (2015) Aree Interne e progetti d'area. Rosenberg & Sellier

Menoni S (2014) Urbanistica e rischio sismico: appunti per uno stato dell'arte a livello internazionale. 154:74–82

Menoni S (2017) Ricostruzione post-sisma nel Centro Italia: la sintesi urbanistica tra esigenze sociali, aspetti ambientali e fisici, la prevenzione dei rischi. Urban Inf 272:763–767

Nonaka I (2008) The knowledge-creating company. Harvard Business Review Press

OECD (2010) OECD reviews of risk management policies: Italy 2010. OECD Publishing

Oliva F (2014) La difficile ricostruzione dell'Aquila. Urbanistica 154:39–52

Olivieri M (2004) Dalla prevenzione edilizia alla prevenzione urbanistica. Urban, Quad, p 44

Olivieri M (2013) Dalla SUM alla CLE: strategie a confronto per la sicurezza degli insediamenti. Urban Doss 130:34–37

Paleari S (2018) Natural disasters in Italy: do we invest enough in risk prevention and mitigation? Int J Environ Stud 75:673–687

Pearce L (2003) Disaster management and community planning, and public participation: how to achieve sustainable hazard mitigation. Nat Hazards 28:211–228. https://doi.org/10.1023/A:1022917721797

Polanyi M (2009) The tacit dimension. University of Chicago Press

Prenger-Berninghoff K, Cortes VJ, Sprague T et al (2014) The connection between long-term and short-term risk management strategies for flood and landslide hazards: examples from land-use planning and emergency management in four European case studies. Nat Hazards Earth Syst Sci 14:3261–3278

Schon DA (1982) Some of what a planner knows a case study of knowing-in-practice. J Am Plann Assoc 48:351–364

Wenger E (2000) Communities of practice and social learning systems. Organization 7:225–246

Zaucha J, Davoudi S, Slob A et al (2016) State-of-the-lagoon reports as vehicles of cross-disciplinary integration. Integr Environ Assess Manag 12:690–700

Chapter 6
Managing Uncertainty: Remarks for Future Policies and Practices

Abstract The chapter summarises the various elements of the changing period we are live in, giving some reflections regarding how to deal with the increasing risks. In the current context characterised by uncertainty, complexity and interdependencies, the concept of resilience has come to prominence in understanding and managing complex systems and its success has enhanced the interest of several research fields. However, the popularity of resilience thinking has not brought any conceptual clarity and, especially, any operational paradigm. For policy makers and practitioners, it is often not clear how to translate into practice the notions of resilience thinking. This chapter identifies some key implications of resilience thinking for management of hazards.

Keywords Disaster risk management · Uncertainty · Resilience · Practices · Knowledge · Learning process

6.1 The Promise of Resilience

Over the last decades, there has been an increase of uncertainty in the context of world "risk society" (Beck 1996, 2009) where society is becoming more interdependent and more complex, and consequently, more vulnerable to new threats and risks associated (see Chap. 2).

There is a global recognition that more concerted efforts are needed to reduce the numbers and effects of natural and anthropogenic disasters. Since the 1990s with the proclamation by the United Nations (UN) of the *International Decade for Natural Disaster Reduction,* which was followed by the International Strategy for Disaster Reduction in 2000, there has been a growing interest in the international agenda on disaster reduction and planning for anticipating and mitigating the impacts of catastrophic events. Subsequently, long-term agreements have worked for a holistic approach that integrates multiple levels of analysis and intervention; multiple time-scales (prevision, response and reconstruction) and multiple sectors of intervention (Chandler and Coaffee 2017). Thus, resilience has gained substantial success, becoming a key factor for many international initiatives (Bourbeau 2015): the Hyogo

E. Pede, *Planning for Resilience*, SpringerBriefs in Geography,
https://doi.org/10.1007/978-3-030-17262-6_6

Framework for Action 2005–2015 (UNISDR 2005) and its successor instrument, the Sendai Framework for Disaster Risk Reduction 2015–2030 (UNISDR 2015), the Intergovernmental Panel on Climate Change (IPCC 2018) and the United Nations Development Program on human resilience (2011). Undoubtedly, the topic is part of the larger challenge of sustainable development that requires a multidisciplinary or interdisciplinary approach, able to consider environmental, human and technological factors and their interdependence.

However, this application of resilience is hard to be operationalised at the different scales. Some of the obstacles are: (1) the pluralistic meanings and interpretations of resilience; (2) the identification of resilience of *what to what*.

6.2 The Meaning of Resilience

While the resilience concept is gaining prominence in different fields, its conceptualisations leave fuzziness and ambiguity (Restemeyer et al. 2018). Chapter 3 has debated the diffusion and the multiple definitions of resilience. Its first use in ecological literature was related to the ability of a system to *bounce back* to an equilibrium point after disturbance. In this meaning, resilience was related to persistence and the capacity to absorb the change of the systems and was called engineering resilience (Holling 1973). Subsequently, the conceptualisation of resilience was influenced by the theory on complex adaptive systems and has been defined as the magnitude of disturbance that can be absorbed before the system changes its structure (Gunderson and Holling 2002; Walker et al. 2004). This interpretation—that was born to describe important changes in ecosystem dynamics, focuses on three characteristics (Brand and Jax 2007):

- the amount of disturbance that the system can absorb;
- the capacity for self-organisation of the system;
- the ability of the system for learning and adaptation.

Subsequently, the concept has been applied to social systems in which the focus concerns communities; but the meaning that has had wider diffusion is resilience with respect to the social-ecological system in which the human system and the natural system are treated together (Folke et al. 2002; Walker et al. 2004). This conceptualisation has brought a new narrative, especially in governmental thinking (Restemeyer et al. 2018) and its application has become very popular in geography and environmental studies and particularly in disaster management, climate adaptation, regional economic development and strategic planning.

Several theoretical elements contribute to the implementation of social-ecological resilience in urban systems:

- humans as part of the ecosystems;
- the urban system as a complex system because of the sum of several subsystems in interaction;

- the ecological, social and economic processes as adaptive cycles that permit adjustment and self-organisation of the systems to find *new orders*.

In social-ecological systems, resilience has its core in the capacity to cope with, adapt to and shape change. Therefore, social-ecological resilience analyses how ecosystems are structured and behave and how institutions and the individuals associated with them are organised and act. Concerning the adaptive cycle in particular, the resilience concept focuses attention on the function of processes and dynamics where individuals, groups, and institutions play a crucial role. Social-ecological resilience is linked to the necessity to learn how to manage uncertainty through change rather than simply to react; the *bounce back* quality, in fact, is undesirable because it will go back to the pre-disaster conditions of vulnerability while the system should evolve in a less vulnerable one.

In this scenario, two main issues emerge: (a) *panarchy*: no system can be understood by focusing at a single scale. All systems function at multiple scales with cross scale interactions, (b) individuals, groups and institutions are responsible for the interactions. In this way, resilience implies a change in policy design, because it highlights the need for flexibility, moving from being rationalist or bureaucratic to becoming more adaptable, collaborative and open to find alternative paths in facing problems. The resilience concept considers dynamism as an intrinsic factor of how the system acts and for this reason it draws an alternative perspective of planning, as more dynamic, fluid and interpretive—because it is partially structured by shared procedures and norms—that are adaptable to changes or uncertainty and help manage them.

6.3 To Operationalise Resilience Through Knowledge

The increasing environmental problems and the many interdependencies with the human activities, the diversity of stakeholder interests and, consequently, the many different view on the risk consequences make resilience an attractive perspective in risk cycle management (see Chap. 1). Studying resilience in emergency planning has several elements of interest: (a) it has to deal with uncertainty; (b) it covers all the disaster life cycle, from prevision to recovery; (c) it bases on a well-defined sharing of power and responsibility among actors. Risk management is recognised as a cross-scale issue including a pluralism of professions working at different levels along different times (Smith 2009, see Chap. 4). The interaction between scales is intrinsic both in theory and in practice. Nevertheless, in the field of practices, the tools (norms, procedures, plans) that should support interaction suffer from several limits that mostly affect the co-management among different actors and, consequently, the knowledge system. Regarding disaster risk reduction, it seems that the knowledge available is not applied or not used effectively to reverse the upward trend in catastrophic events (Weichselgartner and Pigeon 2015). Most of the time, interaction underestimates the need of a common significant framework among actors with different backgrounds, as well as there being difficulties in having feedback loops between different levels. This has raised questions about the effective use of knowledge.

Generally, the challenge of flexibility that the changing period we live in requiring policy, sets also a challenge of dynamism in the knowledge systems and learning processes that support planning, and in particular, risk management. We need a new approach that is aimed at the integration of diverse disciplines knowledge, stakeholder integration, collaborative knowledge production and inclusion of tacit knowledge. Despite the growing uncertainty, decisions have to be made and knowledge is the basis for making it. There is an international recognition of this crucial role; the Sendai Framework for Disaster Risk Reduction 2015–2030 addresses knowledge-related issues and highlights the critical role of knowledge in disaster risk reduction and the importance of its co-production and sharing. However, implementation gaps between research and practice persist and the debate on how to operationalise resilience through an effective use of knowledge is ongoing.

As Sect. 4.5 showed, there are different qualitative levels of understanding: data, information, knowledge. Knowledge is created by organising and by interpreting information. Knowledge is dynamic and is built through social interaction and experience. It can come from different sources (databases, less documents, co-production) or it can have different forms (explicit or tacit) (Cash et al. 2006; Polanyi 2009). Usually, knowledge is scattered among various actors with limited coordination, sharing and coherence. Therefore, the core challenges concern knowledge production processes. Working towards a common vocabulary is considered to be crucial for bridging gaps between the different knowledge producers or users. If knowledge production and learning have become central issues to deal with uncertainty and complexity, can they be the pathway for increasing the resilience of the systems?

6.4 Resilience Is a Long Way

The Umbria region (Italy), over the years, has worked to ensure a more effective use of knowledge in facing seismic risk (see Chap. 5). The case illustrates how the integration and the coproduction of knowledge can trigger a process of transformation of planning practice towards a resilience prospective.

The Umbria experience started with the reshaping of the data management system of emergency planning, but the results have also had consequences on spatial planning and governance over the course of time. At the beginning, the resilience aim was not explicit in the project. Also because of the popularity of the term, the concept has started to be used in the process only in recent years. However, resilience thinking could be a useful analytical framework for understand some key elements of the case study which have determinate its success.

The willingness of the individuals to cooperate emerges as essential. A reflexive approach and high degree of social learning—main components of social-ecological resilience (see Sect. 3.3)—are crucial for creating a constructive dialogue across disciplines able to generate new and useful insights for the risk management. Constructing an effective collaborative process is also a matter of building trust between the parties and social capital in general. Only, the creation of communities of practice (Wenger 2000, see Sects. 4.10 and 5.5) can enhance the knowledge system, increasing the shareable know-how, because it incentives the union of tacit knowledge and practical experience, otherwise unexpressed. Likewise, the creation of formal/informal communities of practice clarifies and makes the different roles and responsibilities explicit, due to stronger accountability and self-awareness. Social capital is important because it triggers a learning-loop. Learning-by-doing has introduced new actors into the practical example, in order to ensure the success of the activities.

Technology can be a helpful tool for supporting the development in information and knowledge management, but it is not the solution (see Sect. 4.9). Initially, the Umbria experience was driven by the technological push of the information technologies and the spatial data infrastructures; however, in short time, it was clear that technology only allows knowledge to be more accessible to people but it needs stakeholder engagement and common vocabulary to be effective (see Sects. 5.4, 5.5). The benefits of using technology concern the possibility of having a knowledge codification through rigid definition of the information. In the same way, it allows to add or simplify complexity according to the scales of analysis, with a major flexibility according to different situations. Nevertheless, codification presupposes that actors have agreed on the meanings and the management rights. Only an extensive deliberation can avoid misunderstanding.

Finally, citizens can play a new important role in the resilience approach: they can have a proactive role and there is growing interest to be involved from the base.

On the contrary, the case has also shown the limit of the resilience approach based on knowledge sharing and learning processes. Firstly, building a solid knowledge system concerning both a database and strong actors' network requires time; time for collecting and building a database, as well as time for accountability and building trust with all stakeholders involved. Stakeholders and actors have been working on their own system for at least ten years and in the future a number of improvements must be made (data updating and enlargement of database). Thus, even if sharing knowledge is a desirable approach, it cannot be the only one. Nowadays, territories are exposed to increasing pressures, in many cases there are territories suffering permanent hazards that need solutions in shorter times. Building a solid knowledge system able to produce self-learning processes can be a long-term strategy, but it is not acceptable in the short term.

Lastly, also the citizens' involvement shows some limits. Generally, citizens can be mostly involved to give or to be educated on how to handle risk. It is much harder to gather information in terms of data quality, accuracy and integrity. However, it is a field in continuous expansion, whereas crowd-sourced spatial data has just started. However, the Umbria case demonstrates the potential of using data collected from people in small towns or remote areas. Nevertheless, another limit of citizens' involvement concerns how to attract their attention, especially in those areas characterised by hazards with a long-time return.

References

Beck U (1996) Risk society and the provident state. In: Risk, environment and modernity: towards a new ecology. SAGE Publications, London

Beck U (2009) Critical theory of world risk society: a cosmopolitan vision. Constellations 16:3–22

Bourbeau P (2015) Resilience and international politics: Premises, debates, agenda. Int Stud Rev 17:374–395

Brand FS, Jax K (2007) Focusing the meaning (s) of resilience: resilience as a descriptive concept and a boundary object. Ecol Soc 12

Cash DW, Adger WN, Berkes F et al (2006) Scale and cross-scale dynamics: governance and information in a multilevel world. Ecol Soc 11:8

Chandler D, Coaffee J (2017) The Routledge handbook of international resilience. Routledge, New York

Folke C, Carpenter S, Elmqvist T et al (2002) Resilience and sustainable development: building adaptive capacity in a world of transformations. AMBIO: A J Hum Environ 31:437–440

Gunderson LH, Holling CS (2002) Panarchy: understanding transformations in systems of humans and nature. Island, Washington

Holling CS (1973) Resilience and stability of ecological systems. Annu Rev Ecol Syst 1–23

IPCC (2018) Global Warning of 1.5 °C (SR15). Summary for Policymakers

Polanyi M (2009) The tacit dimension. University of Chicago Press

Restemeyer B, van den Brink M, Woltjer J (2018) Resilience unpacked–framing of 'uncertainty'and 'adaptability'in long-term flood risk management strategies for London and Rotterdam. Eur Plan Stud 1–21

Smith K (2009) Environmental hazards: assessing risk and reducing disaster, 5°edn. Routledge, New York

UNDP (2011) Towards human resilience: sustaining MDG program in an age of economic uncertainty

UNISDR (2005) Hyogo framework for action 2005-2015. ISDR International strategy for disaster reduction; building the resilience of nations and communities to disasters

UNISDR (2015) Sendai framework for disaster risk reduction 2015–2030. In 3rd United Nations World Conference on DRR

Walker B, Holling CS, Carpenter SR, Kinzig A (2004) Resilience, adaptability and transformability in social–ecological systems. Ecol Soc 9:5

Weichselgartner J, Pigeon P (2015) The role of knowledge in disaster risk reduction. Int J Disaster Risk Sci 6:107–116

Wenger E (2000) Communities of practice and social learning systems. Organization 7:225–246

Index

E. Pede, *Planning for Resilience*, SpringerBriefs in Geography,
https://doi.org/10.1007/978-3-030-17262-6

P

R

S

T

U

The manufacturer's authorised representative in the EU is Springer
Nature Customer Service Centre GmbH, Europaplatz 3, 69115 Heidelberg,
Germany. If you have any concerns regarding our products, please
contact ProductSafety@springernature.com

Printed and bound by CPI Group (UK) Ltd, Croydon, CR0 4YY
05/10/2026
02228151-0003